The Electric Vehicle Revolution

The Electric Vehicle Revolution

Five Visionaries Leading the Charge

Kenneth K. Boyer

ROWMAN & LITTLEFIELD
Lanham • Boulder • New York • London

Published by Rowman & Littlefield
An imprint of The Rowman & Littlefield Publishing Group, Inc.
4501 Forbes Boulevard, Suite 200, Lanham, Maryland 20706
www.rowman.com

86-90 Paul Street, London EC2A 4NE

British Library Cataloguing in Publication Information Available

Library of Congress Cataloging-in-Publication Data

Names: Boyer, Kenneth Karel, 1967- author.
Title: The electric vehicle revolution : five visionaries leading the charge / Kenneth K. Boyer.
Description: Lanham : Rowman & Littlefield Publishers, [2024] | Includes bibliographical references and index.
Identifiers: LCCN 2023059112 (print) | LCCN 2023059113 (ebook) | ISBN 9781538190746 (cloth) | ISBN 9781538190753 (ebook)
Subjects: LCSH: Electric automobiles--History. | Electric vehicles--History.
Classification: LCC TL220 .B679 2024 (print) | LCC TL220 (ebook) | DDC 629.22/9309--dc23/eng/20240223
LC record available at https://lccn.loc.gov/2023059112
LC ebook record available at https://lccn.loc.gov/2023059113

∞™ The paper used in this publication meets the minimum requirements of American National Standard for Information Sciences—Permanence of Paper for Printed Library Materials, ANSI/NISO Z39.48-1992.

Contents

Acknowledgments

The short successes that can be gained in a brief time and without difficulty, are not worth much.

—Henry Ford

This book has been a challenging and exciting journey of discovery and inquiry. So many people have generously shared their time and knowledge that I hope, but doubt, that I have remembered everyone. A million thanks, first and foremost, go to my family, who have been supportive while limiting the eye rolls when hearing yet another fact or story about EVs. My wife, Leigh, children, and bonus children Rob, Jamie, Rachel, Graham, Justine, Julia, and grandchildren Camila and Vince.

Expanding the circle, my agent and partner in this pursuit Herb Schaffner has been a supportive cheerleader, editor and confidante throughout the writing and development. My editor Deni Remsberg has been amazing in holding my feet to the fire, helping me accurately check sources, fine tuning the final product, and reaching for the finish line. Thank you to Deni and Herb.

Academic colleagues were foundational in the early stages of imagining and visioning. Longtime friend and coauthor, Rohit Verma, now dean of the Darla Moore School of Business, University of South Carolina, and his brilliant, lovely wife, Amita, who first invited me to visit VinUniversity in Hanoi as part of a sabbatical semester. This in turn led to an awareness and fascination with the birth and growth of VinFast. Dean Marianne Lewis of the Lindner College of Business, University of Cincinnati, has been a steadfast consultant and supporter. Brian Mittendorf,

Ohio State University, read early drafts and listened patiently on long walks with a cup of coffee.

In Vietnam, Michael Dixon allowed Leigh and me to join a tour with his Utah State students in which we had the opportunity to tour Vinfast's manufacturing facility in Haiphong. Michael Johnson and James DeLuca generously shared their time to discuss their efforts to build VinFast from the ground up.

In Nebraska, Rob Hanson, Dan Levy and Austin Burk were extraordinarily generous in discussing the origin story of Monolith and helping to place its ambitions in the larger context of the automotive supply chain. Similarly, Jessica Hogan was amazing in framing Bolder Industries efforts in the circular economy for tires. Giorgio Rizzoni, Ohio State University, provided a spectacular overview of electric vehicles and their development over the years relative to and alongside ICE vehicles. Giorgio was also extraordinarily gracious in sharing the history of OSU and Honda partnerships over the past four decades.

So many members of the Suppliers Partnership for the Environment also lit the fire of curiosity and engagement in my head. Alissa Yakali of Honda is responsible for introducing me to this amazing group that has been striving for over two decades. Steve Hellem, Kellen Mahoney, and Bob Crawford are all champions of SP for E and welcomed me into the group with open arms.

Many of the DHL's worldwide group of experts and influencers were invaluable on this journey. This includes Florian Schwarz and the many organizers of the Era of Sustainable Logistics conference held in Valencia, Spain, in April 2023. Other DHL partners include Fathi Tiatli, Prerit Mishra, Toby Groom, and Charlene Rudolph.

Many, many others generously chimed in to supplement my archival research and reading. Mitchell Bilo, OSU c/o 2023.5 was an amazing resource in running down research leads and formatting references. In addition, this list includes:

- Bridget Grewal, director, Packaging Continuous Improvement, Magna International
- Jack Ewing, *New York Times* auto correspondent

- Kevin Krolicki and Ben Klayman, Reuters
- Jonathan Bridges, MD, Automotive, Steel and White Goods, JobsOhio
- Mark Williams, *Columbus Dispatch*
- Elowat Depicker, VP Commercial and Corporate Development, Li-Cycle

Finally, thanks to you dear reader for taking some time to read!

Preface

I am seeking to tell a story of the automotive industry in transition, supported by detailed research and analysis of the available research and financial data. I've been a car enthusiast since my father took me to soccer practices in the 1970s in his yellow Chevrolet Corvette Stingray convertible. That car was the epitome of 1960s automotive technology with a standard engine packing 250 horses of power. Fantastically fun to drive and ride in, but amazingly greedy with gasoline—getting about thirteen to fifteen miles per gallon. Of course gasoline was inexpensive and widely available in the United States in the 1960s when GM produced that car. And we didn't yet know the extent of the harmful impacts of the extraction and burning of fossil fuels on our natural environment.

What we did know then was that gasoline and its extraction and refining were smelly and messy, thus people with the means generally chose to live a distance from oil refineries and gasoline stations. In fact, until 1964 customers did not pump their own gasoline, instead an employee of the station would pump it, check the customer's oil, and clean the windshield. Then John Roscoe flipped a switch at a convenience store in Westminster, Colorado, on the north side of Denver to turn on the first self-service gasoline pump. The store only sold 124 gallons of gasoline that day to roughly a dozen customers, but the way people refueled their cars began to change.[1] Today's car drivers almost all pump their own gas, except in New Jersey and Oregon where self service is prohibited. A journey of a thousand miles starts with the first step.

This brings me to why I wrote this book. In addition to believing that electric cars provide a better driver experience than gasoline-powered cars, I also believe the adoption of electric vehicles is one of the best ways to reduce carbon emissions worldwide. But transitioning from gasoline to electricity as a power source requires a transformation of epic proportions. It requires rebuilding society through investments, from mining to production and assembly to the underlying energy grid that powers our daily lives.

My hope is that readers of the book will identify with some of the people profiled while recognizing that it is completely human to innovate and try new things, only to discover over time that any new product or process inevitably has both positive and negative features. As humans advance and gain new knowledge, we hopefully work to mitigate the negatives or unintended consequences of new technologies or systems of living. This book is intended to help readers increase their understanding of the pluses and minuses of electric vehicles, as well as the challenges and opportunities associated with the transformation process.

The book is written so that it can be read in two different ways. The chapters are laid out in a manner that is intended for cover-to-cover reading. At the same time, it is challenging to provide more than a superficial explanation of many of the technical and supply chain challenges involved with changing power sources, packaging, tires, and the overall supply network in a purely linear manner. Thus, the book is organized around the introductory chapter, followed by a strategic analysis of key factors that must be mastered for success. Chapter 2 is titled, pun intended, "Gears of Change." The second section profiles four auto manufacturers driving the industry: Tesla, Vinfast, General Motors, and Honda. The third section focuses on two technical areas where the industry (both the auto manufacturers and their supply network) are pushing for both environmental and societal change: "Power Sources" and "Tires." The last chapter, "Driving It Home," seeks to pull the story together and provides forecasts regarding the chances for success for the individuals and companies described throughout the book. Alternatively to a straight read through, readers also are encouraged to pick and choose the chapters they find most interesting, as my intention is that any of these chapters can stand on its own as a story and analysis.

Finally, another note on focus for the book. The primary focus on American car makers in America is intentional, not to ignore the huge contributions of Chinese and European car makers. Rather this move was to preserve my own sanity and time. The electric vehicle revolution is moving so quickly that this choice was made primarily to have a chance of keeping up with the quicksand of new developments. In fact, as this book goes to press (November 15, 2023) current estimates are that 24 percent of the vehicles sold in China in Q3, 17 percent of the vehicles in Europe, and about 8 percent of American vehicles are pure electric. Apologies for not covering the world—my hope is you find the stories and data reported in this book interesting and consider doing a little of your own research.

Chapter 1

A Tale of Automotive Visionaries

Jim DeLuca began his career near his New Jersey hometown at General Motors' since-shuttered Linden assembly plant. In 1979, Jim was a fresh high school graduate who had enrolled in the co-op program at General Motors Institute in Flint, Michigan. The Linden plant was originally opened in 1937 to build Buick, Pontiac, Cadillac, and Oldsmobile vehicles, switching to production of F4F Wildcat fighter planes for the US military during World War II. The plant returned to automobile production following the war and by the late 1970s was producing luxury models from Buick and Cadillac. Young Jim entered the industry at a time when Japanese manufacturers including Honda and Toyota were coming to America and disrupting Detroit's Big 3. By September 1991 GM idled the facility to re-tool for production of trucks and SUVs, with thousands of workers receiving severance packages and the workers still at the plant not pleased to see their livelihoods inexorably declining. The plant announced plans to permanently shut the Linden plant in February 2002. A white 2005 Blazer was the last vehicle to leave the line on April 20, 2005. The death and closure of the Linden plant is one sad tale among hundreds in the automobile industry.[1]

Flashing forward to 1984, Jim is preparing to graduate from GMI (later renamed Kettering University) with a bachelor's of science in electrical engineering. Very likely Jim crossed paths with Mary Teresa Makela, who graduated with the same degree from GMI in 1985. The daughter of a die maker who labored at GM for thirty-nine years, today she is better known by her married name, Mary Barra. Both veterans of

GM's many upheavals, Jim and Mary moved and rose in the same circles over the ensuing four decades, with assignments and responsibilities that saw them establish themselves as stars of Detroit. During the fifty-six years that GMI was directly supported by General Motors, 96 percent of the graduates went on to work with their sponsoring unit, with three alums rising through the ranks to become CEO.

In February 2014, Jim was appointed executive vice president of Global Manufacturing. Responsible for GM's manufacturing, labor, and operations worldwide, DeLuca bore responsibility for one hundred ninety thousand employees at 171 facilities in thirty-one countries. In this hugely challenging, taxing, and dynamic role, he reported to an old friend as Mary Barra had been named on January 15, 2014, as the first female CEO of a major automotive manufacturer. The business relationship between these two automotive visionaries, a Prince and the Queen, massively influenced our world, both through GM and Jim's future roles.

At the time Jim DeLuca and Mary Barra were coming of age and laying the foundations for their ascendancy in the automotive industry, another pair of future giants were nurtured and raised in Vietnam and South Africa. The Viet Cong forces and the North Vietnamese People's Army of Vietnam launched the Tet Offensive, a shortened name for the Lunar New Year festival, on January 30, 1968. The North Vietnamese government surprised American war planners with eighty thousand Viet Cong troops striking one hundred towns and cities. The hope in Hanoi was that the offensive would trigger a popular uprising and collapse of the South Vietnamese government. The Tet Offensive proceeded through three phases, not ending until late 1968.

In America, the Tet Offensive shocked the public with its massive casualties on both sides. The administration of President Johnson sought negotiations to end the war, but Richard Nixon derailed the talks in in a secret agreement with President Nguyen Van Thieu of South Vietnam. This notoriously underreported self-dealing scheme by Nixon, who planned to run for president in the 1968 election, resulted in years of more death and destruction. Following the Tet Offensive, desertion rates among US forces quadrupled, while enrollment in ROTC declined from almost two hundred thousand to an all-time low of thirty-three thousand in 1974. Essentially US morale and the will to fight collapsed as described by historian Shelby Stanton,

> In the last years of the Army's retreat, its remaining forces were relegated to static security. The American Army's decline was readily apparent in this final stage. Racial incidents, drug abuse, combat disobedience, and crime reflected growing idleness, resentment, and frustration . . . the fatal handicaps of faulty campaign strategy, incomplete wartime preparation, and the tardy, superficial attempts at Vietnamization. An entire American army was sacrificed on the battlefield of Vietnam.[2]

By anyone's estimate the Vietnam War's impact was profoundly awful. In the United States, the war shaped a generation that followed—from the baby boomer generation that fought, tried to evade, and protested the war to Generation X, which came of age in the first war that the United States had lost. A national trauma for the United States, certainly, yet more accurately, an existential one for Vietnam. A 2008 study by the *British Medical Journal* estimated the total number of deaths from 1965 to 1984 in Vietnam as over 2.5 million, with approximately 90 percent being either Vietnamese military or civilians.[3]

Born in 1968 to a family with a father that served in the air force at the height of the Tet Offensive, Pham Nat Vuong grew up desperately poor in Hanoi. Blessed with inherent mathematical ability and nurtured by loving parents, "Vuong Vin" was able to surf a wave with fellow Vietnamese students who were offered the chance to study in the Soviet Bloc. His mathematical talent provided an avenue to study at a geological institute in Moscow in the late 1980s, when Jim DeLuca and Mary Barra were launching their careers with GM. Following graduation, Vuong moved to Kharkiv in Ukraine, where he at first started a Vietnamese restaurant, soon investing in a production line to produce rice noodles at scale. He built Mivina into Ukraine's most popular noodle brand. After selling Mivina to Nestle for $150 million, he returned to Vietnam, where he built a conglomerate. In 2012, Vuong Vin merged two entities, VinPearl, his resort business, and VinCom, his real estate business, creating VinGroup. Forbes declared him Vietnam's first billionaire: "Pham's story personifies the post-Vietnam war story in this nation, a capitalist achievement in a country that remains, nominally, communist."[4]

The distance from Hanoi to Pretoria, South Africa is almost ten thousand kilometers. In Pretoria on June 28, 1971, Elon Musk was born to Maye Musk, nee Haldeman, a model from Saskatchewan, Canada, and Errol Musk, a South African electromechanical engineer. Slightly better known worldwide than Vuong Vin, Elon Musk, the mercurial CEO of Tesla was worth an estimated $175 billion on February 3, 2023, making him the second-wealthiest person in the world. This despite a run of controversies with his acquisition of Twitter that have generated disparaging headlines, including "Elon Musk's Twitter Tantrum Needs a Time Out" and distracted his focus on Tesla. Even worse for Tesla customers and investors, Musk's Twitter trials have focused a plethora of negative press on him and his highly identified status as an electric vehicle (EV) visionary and CEO. At the time of this writing Tesla had peaked at a value of $1.2 trillion in November 2021—more than the next ten automobile manufacturers combined. By May 2023, Tesla's market value had fallen to slightly over $500 billion. At either valuation, the market reflects the amazing innovativeness of Musk and the company. Tesla achieves these valuations despite a far lower price/earnings ratio than any other automotive manufacturer. The company brings in lower revenues and produces fewer vehicles than legacy manufacturers including GM, Honda, and Toyota.

Simply put, the market values of Musk and Tesla have proven that a future with greener, cleaner power, namely electricity, is possible. For this future to be fully realized, the automotive industry and policymakers must execute three key principles and actions. First, a *Willingness to Commit* to a more sustainable future for both people and the environment. Second, the companies implementing such radical innovations must *Transform the Supply Network*. In particular, Tesla had not only to build and develop this network but also had to invent large parts of it—namely, the parts dealing with batteries and electrical propulsion systems versus internal combustion engines (ICE). Finally, Tesla needed to demonstrate an *Ability to Profit*. Check, check, and check. Together these *Gears of Change* can drive us toward a greener, more sustainable world. Success in a nutshell. Yet a question remains: Is Tesla a unicorn or a canary in the coal mine/automotive supply network showing that the oxygen for ICE is rapidly depleting?

This book tells a story about the electric vehicle revolution using five main characters. Elon Musk; Mary Barra (CEO of GM); Pham Nat Vuong, founder of VinFast; Jim DeLuca, who has had leading roles at GM, VinFast, and Ceer; and Rob Hanson, cofounder of Monolith. Each of these leaders has a vision for their company and a greener world. All four are seeking to align the Gears of Change and bring a revolution in transportation.

The Automotive Supply Network 2020

In 2021 an estimated 79.1 million motor vehicles were produced worldwide, with an estimated average lifespan of twelve years—indicating there are approximately 900 million motor vehicles on the road today. Most of these vehicles are traditional ICE vehicles, emitting an average of 4.6 metric tons of CO_2 per year per vehicle. All-in passenger vehicles emit over 3.2 billion metric tons of CO_2. The transportation sector worldwide accounts for 29 percent of carbon dioxide emissions. Road vehicles (cars, trucks, motorcycles, buses) account for three-quarters of this amount, thus, automotive applications account for roughly 22 percent of world CO_2 emissions.[5] Will an EV revolution lead the world to a better natural environment? Further, will this revolution occur in a timely manner? These are the central questions this book examines.

The leading car manufacturing firms have essentially all pledged to transition from traditional CO_2-emitting ICE vehicles to battery-powered electric vehicles (EVs). Tesla has led the way to date, recently producing its four-millionth car since its first sale in 2008. General Motors has committed to carbon neutrality in its global products and operations by 2040 as well as eliminating tailpipe emissions in new light-duty vehicles by 2035. Ford announced in February 2021 that it would invest $22 billion in electric vehicles through 2025. Stellantis (parent company to Chrysler, Fiat, Dodge, Peugeot, Citroen, and several others) has committed to net zero emissions by 2038. Toyota is committed to carbon neutrality worldwide by 2050 and North America by 2035. Lastly, Honda has committed to carbon neutrality and zero traffic collision fatalities by 2050. These are ambitious and noble goals, but they will take a supreme transformation to achieve. Without a supreme transformation, these are empty promises, or what is commonly labeled "greenwashing."

This book examines the prospects and plans for that transformation, starting with the stories of several automotive giants. Each of our protagonists has a vision for transforming the world of transportation into a greener, more environment-friendly industry while also earning profits in our largely capitalist world. In addition to the palpable switch from gasoline to alternative power sources, numerous other challenges/opportunities are associated with environmental and social sustainability. First, while there is a rush to electrify vehicles with lithium-ion batteries, it is not certain that this is the optimal power source in all cases. Thus, chapter 8 will examine efforts to use hydrogen as a power source, particularly in heavy vehicles, including trucking and railroads. Second, there are numerous social sustainability issues. For example, many battery materials such as cobalt, lithium and nickel are largely controlled by Chinese companies, including Zhejiang Huayou Cobalt Co. and Tsingshan Holding Group Co., the world's largest producer of nickel. These companies have been accused of ethically dubious employment practices in Africa and Southeast Asia, with US senator Rubio writing in April 2022:

> As you may know, Huayou has been credibly implicated in the forced labor and human trafficking of child laborers in its cobalt mines. Meanwhile, Tsingshan operates lithium and nickel mines in Indonesia, which require the destruction of the rainforest ecosystem, and is considered a major risk to biodiversity.[6]

While social sustainability is a critical leg of the environmental, social, and corporate governance (popularly known as ESG) stool, this book focuses primarily on environmental sustainability challenges and opportunities. In addition to the impending transition from gasoline to electric or hydrogen as a power source, the automotive industry is innovating its environmental practices in several other areas, of which this book will highlight two.

Roughly 2 billion automobile tires are produced worldwide each year, with the most likely disposal point at end of life being a landfill or tire dump. In the 1980s and 1990s, Kuwait developed a business model importing discarded tires from the United States and Europe, accepting a quarter billion tires per year until the practice was banned in 2001. In October 2020, a fire at the Al Sulabiya tire site outside Kuwait City burned through an estimated twenty-five thousand square meters of a 1-million-square-meter tire graveyard (roughly 4%). Estimates are that over a million tires burned; smoke from this catastrophe was visible to the human eye from space.[7]

Creating a tire recycling system would alleviate this environmental disaster, yet it is an enormous challenge. Approximately 10 percent of tires are recycled worldwide, with most being burned for fuel. Better than sitting in a dump collecting water, insects, rats and potentially creating a fire visible from space? Yes, yet far from perfect. Manufacturing tires produces greenhouse gases and carbon dioxide, while driving on the tires creates almost two thousand times as many toxic particle emissions as those from gasoline-powered car tailpipes. Finally, burning the tires for fuel recaptures some energy but releases many other nasty things.

Chapter 9 will focus on this challenge, examining existing problems with the tire production and disposal system. Another way to look at challenges is as an opportunity to improve. This is the core business model of Bolder Industries, which offers "circular solutions for rubber, plastic, and petrochemical supply chains." In practical terms, Bolder says that it can recover 98 percent of the raw materials in used tires while reducing total emissions and making money. Another company, Monolith, operates further upstream in the supply chain. Monolith claims that it's "proprietary methane pyrolysis technology uses renewable electricity to split natural gas into hydrogen and highly valuable solid carbon materials without emitting carbon dioxide." The bold claims of both Bolder Industries and Monolith will be examined, explained, and analyzed in chapter 9. In addition, the free market will naturally test these firms and the others seeking to change the industry in terms of *ability to profit*.

Any examination of batteries for EVs must start with Muskla. Thus, chapter 3, "The Birth and Evolution of Muskla," examines Tesla's birth, growth, and achievements as led by Elon Musk. Note, there are five men credited with founding Tesla, and while Musk has gained the lion's share of wealth and attention, he is far from the sole contributor. At the same time, Elon Musk and Tesla have become almost interchangeable in the public's eye, thus the occasional use of the term *Muskla.* I examine the birth of Tesla and profile the full quintet of Tesla founders responsible for its existence.

Under Elon's leadership as CEO, Tesla grew from sales of twenty-five hundred roadsters in 2008 to becoming the first company to sell 1 million electric vehicles in June 2021. Tesla reached its maximum capitalization to date of $1.23 trillion on November 22, 2021, based on revenues of $53.8 billion. A concise explanation of Tesla's rise would note that it essentially proved the viability of electric vehicles, yet there is much more to the story. Chapter 3 examines the many challenges faced and overcome during the company's evolution. In short, these include, transforming the supply network by building a new car brand using a new propulsion source, namely lithium batteries. The company needed to create a supply network capable of competing with GM, Ford, BMW, Toyota, and Honda, while also finding and delivering a viable solution for powering automobiles. Second, Tesla pioneered the idea that cars could run on electricity and emit far less carbon dioxide. The company didn't pioneer this concept; it resurrected it. Early automobile manufactures at the beginning of the twentieth century offered several models of electric cars. Clearly, these cars did not win the market of the early 1900s. Thus, Muskla had to be willing to commit to this power source. Just as critically, consumers needed to be convinced. Finally, in a capitalist economy, Tesla had to be able to produce and sell EVs profitably.

Chapters 4 and 5 focus on a trio of giants. First the relationship between Jim DeLuca and Mary Barra is examined. Then, chapter 5 examines the birth and growth to date of VinFast and its chairman and founder Pham Nat Vuong. Like Tesla, VinFast seeks to import automotive talent from around the world, first to Vietnam and then to the United States in North Carolina. The challenges are immense, the opportunities are as well. A key difference is the level of nationalistic fervor and support for VinFast, labeled the "Vietnamese national car."

From here, the book will examine the efforts to reimagine an industry icon. Chapter 6 will focus on General Motors' Comeback Queen, Mary Barra, and her leadership of the company into an electric future. The "Comeback Queen" label was bestowed when disaster struck two weeks into her term as CEO in early 2014. She became CEO the same week the US government ended a $50 billion bailout of GM stemming from the 2008–2009 financial crisis. Over a handful of years, GM was able to repay every penny of the loans provided at the height of the crisis. The smooth driving ended there.

In her second week on the job Barra was cruising in her black Cadillac Escalade on her way home when she took a call that rattled her to the core. A senior colleague was calling to inform her that many GM models had faulty ignition switches that had caused at least 124 deaths. Even worse, engineers in charge of the design process had known about the fault—which disabled power steering and airbags—for over a decade. From elation at being elevated to top dog, her emotion "turned to disappointment. It was very difficult." By far, the sacking of the CEO is the most likely outcome in a scandal that results in recall and repair of 30 million vehicles, a cost of $4 billion and federal fines totaling $1 billion. Not Barra.[8]

Leading by empathetic and direct example, she quickly dispelled the insulting moniker bestowed on her as a "lightweight." Barra's promotion to CEO in 2014 made sense to colleagues within the company and the board, yet outside the firm, opinions were less positive. "She has not distinguished herself in any heavy-duty operating role. She's sort of a blank slate," according to an investment banker who had worked with GM.[9] Key moves in steering through the crisis included admitting "unacceptable and disturbing" mistakes, meeting with families of victims, setting up a compensation committee and firing fifteen executives. Her loyal lieutenant during this time? Jim DeLuca was quoted in July 2014 regarding the promotion of Cathy Clegg to VP of North American manufacturing saying, "Cathy's proven track record in key labor and manufacturing roles and her passion for innovation will be an asset to GM and her team. Her demonstrated commitment to employees, and our customers will help ensure our continued focus on safety and quality."[10] Was Clegg's promotion to replace one of the fifteen executives fired for their culpability in the ignition switch recall? My guess is yes.

Following the recall firefight, General Motors, under Barra's leadership, turned its focus to electric vehicles. Clearly the automotive industry is composed of millions of individual contributors. Elon Musk, Pham Nat Vuoung, Jim DeLuca, and Mary Barra humanize this immense industry and help tell the tale of this transformative moment in history.

The final automotive company examined will be Honda in chapter 7. Founded by Soichiro Honda, the company has been the largest manufacturer of motorcycles in the world since 1959, reaching a total aggregate volume of 400 million. In addition, Honda is the world's largest manufacturer of internal combustion engines, with an annual production of 14 million. The company serves as a good comparison with VinFast because it grew in a country (Japan in the 1940s and 1950s) where motorcycle ridership outnumbered that for automobiles, it developed a core competence in ICE, then exported this expertise, building an assembly plant for the Accord in Marysville, Ohio, to penetrate the US market in 1982. By 1989, Honda produced the best-selling car in America, the Accord. In many ways VinFast is seeking to copy Honda's playbook, yet at the same time VinFast is seeking to leapfrog Honda by making the jump from ICE to EV cars. Meanwhile, Honda, with revenues of $129 billion in 2022, is a nervous giant. Somewhat behind in the transition to EV's, the company's mantra as early as 2018 was CASE—Connected, Autonomous, Shared, and Electric—as the leaders forecast the future. Yet a forecast has been called an educated guess, executing and succeeding on that vision is a substantial hurdle. With the four focal car companies outlined, I travel back in time to learn from the history of an early breakout star of the automobile industry.

THE OG OF THE AUTOMOBILE INDUSTRY—HENRY FORD

Henry Ford, born in 1863 on a farm in Michigan to a father who emigrated from Ireland during the potato famine, died in 1947 in a world that he had helped shape and create. Upon finishing eighth grade at a one-room school, Springwells Middle School, he never attended formal schooling again. Widely known as the "inventor" of the moving assembly line, this is not strictly true. Instead, Henry Ford was a brilliant observer of processes and innovator/adapter of methods and technologies pioneered by other men.

Leaving home in 1879 to begin work as an apprentice machinist in Detroit, he recalled that in 1892,

> I completed my first motor car, powered by a two-cylinder four horsepower motor, with a two-and-half-inch bore and a six-inch stroke, which was connected to a countershaft by a belt and then to the rear wheel by a chain. The belt was shifted by a clutch lever to control speeds at 10 or 20 miles per hour, augmented by a throttle. Other features included 28-inch wire bicycle wheels with rubber tires, a foot brake, a 3-gallon gasoline tank, and later, a water jacket around the cylinders for cooling.

That original automobile, or horseless carriage, set Ford on a path that led to employment at the Edison Illuminating Company in Detroit, which he resigned from to found the Detroit Automobile Company on August 5, 1899. With automobiles that were of lower quality and higher price than Ford desired, the company failed and was dissolved in 1901.

Yet Ford kept at it, bringing in another group of investors to form the Ford Motor Company with $28,000 in capital in 1903. To place this in perspective, that amount in 2023 dollars is equivalent to over $900,000. An advertisement for "Pleasure Automobiles a List of Those on the Market" in Frank Leslie's *Popular Monthly* in 1904 contained eighty-eight models of gasoline, steam, and electric carriages. These ranged from a $2,500 Pierce Arrow (about $83,000 in 2023) to Ford's more affordable and less powerful eight-horsepower. The Ford was priced at only $750 (about $25,000 in 2023). *Frank Leslie's Popular Monthly* also contained electric models including the Baker Runabout from the Baker Manufacturing Company of Cleveland also priced at $750. At the time, automobiles were well out of the reach of all but very wealthy people.

Throughout these years, Ford kept working, experimenting, and innovating. The Model T debuted in October 1908 at only $825. This model was Ford's twentieth, and for a very long time, his last. The car was produced essentially unchanged from 1908 until 1927, when Henry watched the 15-millionth Model T, or Tin Lizzie roll off the assembly line. The innovations that helped make this the most influential automobile in history included ease of repair, mass assembly, and frequent reductions in price. Ford provided a printed manual that helped owners "standardize the method of repairing Ford cars, to ensure continued and satisfactory performance of our product the world over." Explicit and simple repair instructions were critical to developing this new market that did not have readily available gasoline and what used to be called service stations, to repair frequent breakdowns. This manual is today commonly known as the "Model T Bible" to enthusiasts and collectors.[11]

Another key Fordism was implementing the moving assembly line in his Highland Park assembly plant. While many believe Ford invented the moving assembly line, he copied it from the meatpacking industry in Chicago. Ford and his engineers worked relentlessly to improve the production process for the Model T (largely keeping the product or car itself the same). The relentless focus on process improvement resulted in efficiencies that allowed Ford to reduce the price of a Model T to $360 in 1916—the equivalent of $9,800 in 2023. Sales reached 472,000 in 1916, with many of the cars bought by his workers, who he famously began paying $5 per day for a forty-hour week in 1914, enabling middle-income families to afford an automobile.

Before the Model T, automobiles were unaffordable to all but the very rich, and there was a wave of hostility toward horseless carriages. In 1906 the *North American Review* published an article titled "An Appeal to Our Millionaires," writing, "Unfortunately, our millionaires, and especially their idle and degenerate children, have been flaunting their money in the faces of the poor as if actually wishing to provoke them. The rich prefer to buy immense cars which take almost all of a narrow street or road, and to drive them on all streets, narrow or wide, at such speeds as imperils the lives and limbs of everybody in their path." The same publication estimated that more Americans had died in car accidents during the first half of 1906 than had perished in the Spanish-American War.

In 1900, about eight thousand cars were registered in the United States; that number leaped to two hundred thousand by 1908 and almost half a million two years later. Motorists enjoyed the freedom afforded by the ability to drive longer distances, and one of the keys to the Model T's success was its appeal to women. The car was far easier to operate than most of its competitors, and Ford capitalized on this advantage as a company publication boasted, "There is no complex shifting of gears to bother the driver. In fact there is very little machinery about the car—none that a woman cannot understand in a few minutes and learn to control with a little practice." Eighty years before Thelma and Louise left their dreary lives in Arkansas for adventure, Ruth Calkins of Rochester, New York defied attempts to dissuade her from taking a trip without a man along to attend to mechanical challenges. She and three female friends toured the northeastern United States and southern Ontario for a month. Even when the car sank axle-deep into the mud, she and her companions could ease the car out with careful cunning rather than rely on the brute force of a male companion. Ford published a publicity pamphlet titled *The Woman and the Ford* stating, "It has broadened her horizon—increased her pleasures—given new vigor to her body—made neighbors of far-away friends—and multiplied tremendously her range of activity. It is a real weapon in the changing order. More than any other—the Ford is a woman's car."[12]

A Revolution in Automobile Production

Few will argue Henry Ford's key role in history and influence on the world. How will today's giants at Tesla, VinFast, GM, and Honda influence our world?

To illustrate the transformational challenges, consider sales and production figures gleaned from the data aggregation company GlobalData. The largest single car market in the world is China, which overtook the United States for that distinction a few years ago. China has over 220 million registered cars, with the United States coming in with a little over 215 million registered cars. Not only is the China market huge, but it is leading the world in EV sales with the sale of 5.4 million in 2022, or about two-thirds of global sales of 8 million. After China and the United States, Japan and Germany are the next largest markets with roughly 60 million and 40 million registered cars, respectively. US, European, and

Japanese manufacturers—the traditional "world leaders"—have been losing market share in China rapidly, in part due to strong incentives from China's ruling Communist Party. This book will not actively profile any Chinese car manufacturers due to scope issues, yet the success of EVs in China requires some attention to companies such as BYD Auto which competes with Tesla for the title of best-selling electric automaker.[13]

I use registrations instead of annual sales because it emphasizes the total size of the market and the challenge of transitioning an entire industry from ICE to EV. One of the categories of registrations tracked in most developed countries is cars over twelve years old, which is over 20 percent of all automobiles. Consider the ramifications—if every automobile company instantaneously gave up on the ICE descendants of the Tin Lizzie, it would be at least a decade before the fleet of cars on the road even approached being 50 percent EV!

The Rise of Muskla

Elon Musk is a brilliant entrepreneur who has played a significant role in at least a half dozen companies. Founding Zip2 in 1995 with his brother Kimbal and Greg Kouri, the company developed an internet city guide with maps, directions, and yellow pages. The company marketed its services to newspapers—back when they were still printed. The company was sold to Compaq computers in 1999 for $307 million, with Elon receiving $22 million for his 7 percent share. Next, in 1999 Musk cofounded X.com with three other partners. This online financial services and e-mail payment company eventually became PayPal, which eBay acquired for $1.5 billion in stock, with Musk receiving $175.8 million for his 11.72 percent stake. Zip2 and PayPal are software companies with minimal if any tangible product. Musk and Tesla make a very tangible product, with thousands of parts, that is expected to perform near flawlessly for a decade or more. Here I turn to the story of building a radically different car company.

It is clear that while Tesla has proven the viability of EVs, the legacy automakers such as GM, Honda, and Toyota do not intend to allow Tesla to dominate the industry unchallenged. Throughout 2022, hardly a week passed without one company or another announcing a new EV it planned to roll out. Tesla's revenues in 2021 were approximately $54 billion, while it was valued at over $1.2 trillion. For the same fiscal year, Ford Motor (now the fourth-largest car company in the world) had revenues of $127.1 billion and a valuation of $77.3 billion. In other words, with over twice Tesla's revenues, Ford was valued at only about 10 percent of Elon's company. In fact, in November 2021, Tesla's total valuation was higher than the ten largest auto manufacturers combined, including household names like Volkswagen, Toyota, Daimler, Ford, General Motors, Honda, BMW, SAIC (the second-largest Chinese auto manufacturer), Stellantis, and Hyundai.

Chapters 2 and 3 examine the headwinds Tesla confronted beginning in 2022 as the entire automotive industry committed to transitioning to full electric vehicle production over a period of less than two decades. The pledged commitments portend a revolution in economics and emissions that exceeds that of the early twentieth century when Henry Ford's gasoline powered cars began establishing a mass form of transportation that would replace horse-drawn carriages and steam-powered trains. Many around the world argue that the lean production pioneered at Toyota, adopted by Honda and examined in fascinating detail in Womack et. al.'s 1990 classic *The Machine That Changed the World*, published in 1990, represents a second automotive revolution—one that changed production systems and fostered a radical improvement in the quality of cars. Incorporated in 1949, Honda Motor Co. used elements of the Toyota Production System to evolve into the largest motorcycle manufacturer in the world by 1959. Founded by Soichiro Honda, the company built motorcycle and car manufacturing plants in Ohio in the late 1970s and early 1980s. In 2023, Honda will have manufacturing facilities on four continents, including motorcycles and cars in Vietnam. I profile Honda and its current efforts to move into the EV space in chapter 7 because VinFast is imitating its strategic model in many ways, while seeking to leapfrog it by moving rapidly into EV manufacturing.

Lean production certainly played a significant role in Honda's growth and success, yet I don't believe this qualifies as a revolution. Merriam-Webster defines a revolution as "a sudden, radical, or complete change." My research and experience with lean production or operational excellence suggests that the period of 1980 through 2010 was more of an insurrection, which Merriam-Webster defines as "an armed uprising that quickly fails or succeeds." Established structures and power sources (ICE) persisted as the automotive network widely adopted operational excellence, becoming more efficient and increasing the range of available automotive options. Yet there was no need to destroy existing structures. As *The Machine That Changed the World* illustrated, although lean production or operational excellence offered a superior automobile production system across all of the countries and manufacturers, it did not portend a revolution.

Daniel Burnham, the architect and city planner of Chicago, is reported to have said after the Chicago fire of 1909, "Make no small plans for they have no power to stir men's souls." Powerful stuff, yet another of Burnham's quotes is more apropos for this book and Pham Nat Vuong's vision:

> Our city of the future will be without smoke, dust or gasses from manufacturing plants, and the air will therefore be pure. The streets will be as clean as our drawing rooms today. Smoke will be thoroughly consumed, and gases liberated in manufacture will be tanked and burned. Railways will be operated electrically, all building operations will be effectually shut in to prevent the escape of dust, and horses will disappear from the streets. *Out of all these things will come not only commercial economy but bodily health and spiritual joy.*[14]

I emphasized the above words as indelibly appropriate for the theme of this book. Burnham is remembered as a visionary architect who helped build the streetscape of modern America. One of his major roles was as director of works for the 1892–1893 World's Columbian Exposition, commonly called the White City. He also created the Plan of Chicago for rebuilding the city after the Great Fire of 1909. In fact, the current Museum of Science and Industry is housed in the Palace of Fine Arts that Burnham designed in 1893. The architect helped birth modern Chicago, yet few who visit the Windy City would agree the streets are as clean as "our drawing rooms." Visionaries combine immense ambition and less-than-perfect vision. Can Pham Nat Vuong and VinFast have a comparable influence on not only Vietnam but also US and world auto markets? Vuong certainly talks like a man who intends to. In an interview with *Car and Driver*, he said, "Vietnam is not known for being an industrial country. . . . It is not known for manufacturing, but we are developing." Predicting VinFast as a world leader in EV manufacture and sales, he went on, "Maybe not in five years, but in 10? We want to be at the top. Life is short, I cannot be slow."[15]

A man with huge visions and an impatient persona, Vuong and his company announced in March 2022 the acquisition of a vast mega parcel of land in North Carolina and a plan to invest $4 billion to build a manufacturing plant to produce EVs to be sold in North America. This plan is beyond audacious. No Vietnamese automaker has ever produced more than twenty thousand cars in a year. VinFast was able to build an 827-acre manufacturing complex on land reclaimed from the ocean in only twenty-one months and produce and sell five thousand traditional ICE cars in the first quarter of 2020, making it the fifth-best seller in the market behind established companies Hyundai, Toyota, Kia, and Honda.[16] Not bad, but a far cry from market domination and not in the same ballpark as producing and selling an unknown and unproven product in the world's most demanding automobile market.

Estimates indicate that sales of EV cars rose to 5.6 percent of new car sales between April and June 2022 in the United States, while sales run at about 10 percent in Europe and 20 percent in China.[17] Tesla has tracked very closely to Model T production. Figure 1.1 compares Ford Model T production from 1909 to 1927 with data from Tesla worldwide production matched. Note a couple of things. First, the figure omits Tesla data from 2008 to 2013. Second, the data for 2023 for Tesla are a forecast based on the first half-year production. It is eerie how closely the production patterns match! At the same time there are several fundamental differences between Henry and Elon's eras. First is that by 1918 Model T sales represented 50 percent of all automobile sales in the United States, while in 2022 Tesla represented roughly 4 percent of US sales. A second fundamental difference and barrier to transformation is the complexity and scale of the existing global supply network.

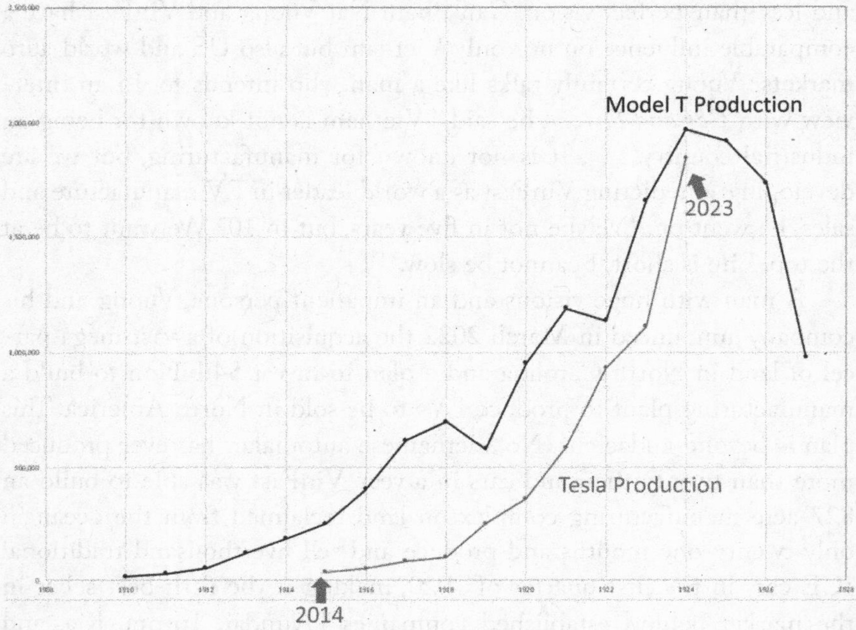

Figure 1.1. Tesla and Model T Production Compared

While the Model T fell in price from $825 (~$25,000 today) in 1909 to $364 (~$5,700 today) in 1923, with sales jumping from 6,389 to over 1.9 million, Tesla prices range from $35,000 to over $100,000 in 2022 dollars. Currently, EV cars are only available to the wealthy portion of the market. In addition, the modern automotive supply network is huge and incredibly complex. Consider two pieces of data. First, a typical car consists of over twenty thousand individual parts assembled by a single OEM, each of which collaborates with hundreds of Tier 1 suppliers, who in turn collaborate with hundreds of Tier 2 suppliers and on and on. Overall, the top twenty OEMS had revenues of over $2 trillion in 2021, roughly equivalent to the aggregate GDP of Italy, the eighth-largest economy in the world.

Returning to Figure 1.1, Tesla has achieved some amazing success, but now that the industry is all in on EVs, it is reasonable to wonder if Tesla will continue to scale? Two clues are offered by the decline in production for the Tin Lizzie. Retired in 1927 due to flagging demand, the

Model T was replaced by the Model A, which while successful, did not reach sales of 1 million cars until February 1929, a full sixteen months after initial production. Ford fumbled the switchover with a five-month period between the last Model T and the first Model A rolling off the line and substantial revenue lost. Offered in four colors rather than the monotone black of the Tin Lizzie, the Model A was Ford's response to General Motors and Alfred Sloan introducing the concept of model years, customer-chosen options and multiple colors. In 1927 General Motors surpassed Ford in car production for the first time, with over 1.4 million produced to Ford's 367,000.[18] By 1953 General Motors dominated the US economy and the minds of elected officials. When Charles Wilson, GM's president and candidate for President Eisenhower's secretary of defense was asked in his Senate confirmation hearings whether he might have a conflict making government decisions that might not be in the interest of GM he had a microphone dropping reply. He said he found it hard to imagine a conflict of interest "because for years I thought what was good for the country was good for General Motors and vice versa." Hence, we see our first giant, Henry Ford, stumbling a bit. Today with competitors focusing on electric vehicles, Tesla's stock value fell by roughly 50 percent from early 2022 to mid-2023. Might our present-day giant, Elon Musk, stumble? Many in the automotive industry are seeking to take him and Muskla on.

May 2019, Vietnam

John Reed, the *Financial Times* Bangkok bureau head, travels from Hanoi to the port of Haiphong, traveling over a causeway to Cat Hai Island. Vietnam's largest company, founded by the country's richest man, Pham Nat Vuoung, was building an integrated production plant on reclaimed seabed. From concept to completion, the project was completed in twenty-one months, an amazing feat compared to historical plant construction projects requiring eight to ten years. The construction was so swift that Google Maps showed Reed standing in seawater in the Gulf of Tonkin.[19] Vuong with the imported knowledge of hundreds of ex-pats from leading automotive sector giants, including tip of the tongue names such as General Motors, BMW, and Bosch, a major automotive supplier, built the VinFast plant in Haiphong and presents it as a robot-filled, state-of-the-art factory.

Yet, there is far more to the story.

Vietnam's agricultural economy was devastated by what Vietnamese call the American War from 1945 to 1975. In 1986 the sixth national congress of the Communist Party of Vietnam passed a set of socialist-oriented market economic reforms. These reforms, commonly referred to as Doi Mos encouraged private ownership of enterprises. Following 8 percent annual GDP growth between 1990 and 1997, the country's GDP per capita stood at only $353, by 2021 this number had increased ten-fold to over $3,700. Over two decades as a foreign correspondent, John Reed has "found Vietnam's communist government to be one of the most business-friendly I have encountered. As in China, the combination of a vibrant business environment with a non-democratic single-party state has helped to create the conditions for an economic take-off." The words are an understatement in this case, with the country exhibiting some of the highest economic growth in the world. Market research group Nielsen estimated that average wages grew by 17 percent and personal disposable income by 29 percent between 2014 and 2018. Great for the average Vietnamese citizen, yet as in many Western economies, the spoils are unequally shared. Nielsen estimates that affluent Vietnamese or US dollar millionaires will have grown 170 percent to 38,600 in the decade leading to 2025. At the top is VinFast's founder, often nicknamed "Vuong Vin," estimated to have a net worth of $7.6 billion.

Born in 1968 to a family with a father that served in the air force at the height of the Vietnam War, Vuong grew up in Hanoi. His mathematical talent provided an avenue to study at a geological institute in Moscow. Following graduation he moved to Kharkiv in the Ukraine. Searching for business opportunities in the disintegrating Soviet bloc, he partnered with another Vietnamese student, Le Viet Lam, to start a restaurant. Recognizing the benefits of scaling production, the two created Technocom, bought a production line from Vietnam and built the company into Mivina, Ukraine's most popular noodle brand. With that venture selling to Nestle for $150 million, Vuong had a capital stake to support Vingroup, which began operations in Vietnam in 2000 and was first listed on the Ho Chi Minh City Stock Exchange in 2007. While Vuong focused on the growth of Vingroup, Le Viet Lam went on to found Sun Group with three fellow Vietnamese who had studied in the Soviet Union. According to Wikipedia, Sun Group employed four thousand and had built seven different entertainment destinations in Vietnam as of 2015.[20]

In the early part of the twenty-first century, Vuong spearheaded the Vinpearl Resort Nha Trang (opened in 2003) and Vincom City Towers in central Hanoi in 2003. After Vincom went public in 2007 it merged with Vinpearl, the luxury resort business to form Vingroup. By 2015 Vuong was listed as the richest person in Vietnam with assets totaling 24.3 trillion Vietnam dong, or approximately $1.2 billion, more than quadruple the assets of the second-richest Vietnamese. The extent of Vingroup and Vuong's influence within Vietnam can hardly be overstated. By the close of 2022, the company's portfolio included business units focused on residential construction (Vinhomes), smartphones and televisions (VinSmart), education (VinSchool and VinUniversity), and, importantly for our purposes, a huge push into automobiles named VinFast. The company reported $4.8 billion in revenue for 2020, representing over 1 percent of Vietnam's entire domestic product. Yet, while Vuong and his team paint an idyllic picture there are many challenges and uncertainties.

The formation of VinFast and the construction of the plant in Haiphong in about twenty-one months appears to be incredibly impressive. At the same time, the carefully curated scene that Vuong wishes to present may be a Potemkin village. News stories and videos suggest a highly automated plant that achieves near perfection and can produce two hundred fifty thousand automobiles per year. Careful examination of a video produced and aired by a North Carolina television station shows a sea of two-wheeled scooters—a wonderful product, yet a far cry from a four-wheel car with an internal combustion engine that will meet the demands of customers in the United States and other developed countries.[21] In August 2022, the company announced it would end the production of ICE vehicles and focus on EV production after producing and selling sixty-five thousand more traditional ICE vehicles in 2020 and 2021.[22] This might be a brilliant, foresighted move of genius that allows VinFast to move quickly down the learning curve and leapfrog entrenched rivals. Alternatively it might be a strategic blunder that deprives the company of a ready source of ongoing revenue to support its vast ambitions.

In March 2022, VinFast announced it had secured $4 billion in funding and $1.2 billion in incentives from the state of North Carolina. Vinfast planned to build a factory complex to manufacture batteries and vehicles on a mega-site in Chatham County, outside Raleigh. In March 2022, this announcement was trumpeted by a tweet from President Biden.[23]

> Today's announcement that the electric vehicle maker VinFast will build an electric vehicle and battery manufacturing facility in North Carolina—$4 billion to create more than 7,000 jobs and hundreds of thousands of electric vehicles and batteries—is the latest example of my economic strategy at work.

Having the president of the United States tweet about a planned manufacturing plant and being backed by major financial institutions Credit Suisse and Citigroup, who had committed to arrange financing of up to $4 billion certainly enhances name recognition,[24] as does the Bloomberg announcement in December 2022 that an IPO and listing on the NASDAQ was in the works.[25] How do we separate hype from reality? Put another way, should I, or you, invest in VinFast and/or consider purchasing one of its cars?

To begin addressing these questions we examine the supply chain mechanics of manufacturing automobiles, which are incredibly complex with over six thousand parts coming together on a single assembly line to produce typically thirty to sixty vehicles per hour. It is magical to watch when it works, yet the logistics and millions of processes are incredibly intricate. Our next key leader is Jim DeLuca who served as CEO of VinFast and later deputy CEO of Vingroup from 2017 to 2021.

A brief review of his profile on LinkedIn reveals the arc of Jim DeLuca's career and hence the reason we label him as a Prince of Detroit. His alma mater, Kettering University, was founded in 1919 as a small trade school in Flint, Michigan, and originally named the School of Automotive Trades. It welcomed both skilled and unskilled factory workers who sought to become engineers, managers, and business executives. The school was so successful that General Motors took over financial support of it and renamed it the General Motors Institute, or GMI. The school offered a cooperative education program where students alternated semesters enrolled in Flint with semesters working in General Motors facilities. GMI provided a pipeline of talent straight to the company and produced at least three CEOs, including the first woman and current CEO, Mary Barra, whom we profile in chapters 4 and 6. In 1974, hurting from the oil embargo crisis and poor financial results, the company issued a directive to cut financial support for GMI by a third. In 1981 a GM task force recommended the company drop GMI within two years. While there were proposals to transition the school under the umbrellas of either University of Michigan or Michigan State University, GM agreed to provide $24.6 million in transition funding and the school was rebranded as Kettering University. Honoring Charles Kettering, holder of 186 patents and the head of research at GM from 1920 to 1947, the new name conveyed the importance of innovation and technological know-how in the automotive industry. Among Kettering's influential inventions were the electrical starting motor, which helped GM catch up to and surpass the Tin Lizzie. In addition, Kettering was a leader in the development of leaded gasoline, which helped reduce the knocking of early automobile engines. Leaded gasoline solved one problem while unintentionally creating a different one, namely, the European Chemicals Agency classifies it as a *substance of very high concern*. Beginning in 1975 the US mandated the use of catalytic converters to reduce emissions, beginning a phaseout of leaded gasoline. This book will examine innovations in the industry at present during this transformational period. As with the discovery, adoption and then phaseout of leaded gasoline, we will see technology that varies widely in terms of success rate, and some of the technologies that initially appear to be positive may later have downsides or limitations revealed.

Following GMs investment in Kettering, the university's leadership expanded its co-op program to work with other leading manufacturing programs and diversify its educational agenda. I am proud to say my own father graduated with a degree in mechanical engineering in 1961. Jim DeLuca graduated from Kettering University with a bachelor's in electrical engineering in 1984 and a master's in manufacturing management in 1987. His future CEO, classmate, friend, and rival, Mary Barra graduated with the same degree in electrical engineering in 1985.

Chapter 2

Gears of Change

In the late nineteenth century and into the first two decades of the twentieth, numerous efforts were made to develop horseless carriages or automobiles. Early efforts were focused on a reliable source of portable power. Steam seemed logical, as it worked for railroads. The state of Wisconsin in 1875 offered a prize of $10,000 (equivalent to $246,758 in 2021) to the first to produce a substitute for the use of horses and other animals. The rules stipulated that the vehicle would have to maintain an average speed of five miles per hour over a two-hundred-mile course. The offer led to the first city-to-city automobile race in the United States, starting on July 16, 1878, in Green Bay, Wisconsin, and ending in Madison, Wisconsin. While seven vehicles were registered, only two started the race: entries from Green Bay and Oshkosh, both of which were powered by steam. The vehicle from Green Bay was faster but broke down before completing the race. The Oshkosh vehicle, engineered by twin brothers Francis and Freelan Stanley, finished the course in thirty-three hours and twenty-seven minutes and posted an average speed of six miles per hour. In 1879, the Wisconsin legislature awarded half the prize. The Stanley Motor Carriage Company continued to produce steam vehicles into the 1920s. Other early efforts included electric- and gasoline-powered automobiles. One of the longest surviving electric cars was made by the Detroit Electric Car Company through 1929. During this period there were well over one hundred companies competing to develop a viable and profitable automobile. Most failed. A few giants including Henry Ford, William Durant, Alfred P. Sloan, Ransom Olds, and Walter Chrysler forged paths that led them to great success and huge fortunes. In Germany, Gottlieb Wilhelm Daimler and Carl Benz had competing claims on having produced the first viable gasoline-powered car. This fierce rivalry prevented any type of cooperation until Daimler and Benz merged in 1926, with Mr. Daimler having died in 1900 and Mr. Benz reaching the age of eighty-two before his passing in 1929. Another catalyst for the merger was the entry of General Motors into the German market and the threat posed by the world's then largest car manufacturer.

Today the automotive industry is in the early stages of a world-altering bet on a future with clean, renewable energy sources. Similar to the early twentieth century, there is a great deal of uncertainty regarding the best technologies and approaches for powering and selling automobiles. Furthermore there almost certainly will be similar losers and winners with some individuals, companies, states, and countries gaining higher market share and wealth. The transformative period we are in has huge stakes and consequences for everyone on the planet. This chapter focuses on a trilogy of interrelated concepts that are foundational for the success of this transformation. My hope is that this chapter begins to provide a framework for individuals, companies, and governments to develop a strategy, make operational decisions, and develop public policies that contribute to a greener, more socially sustainable economy worldwide.

To achieve a more sustainable automotive industry requires an interlocking combination of willingness to commit, ability to profit, and a transformation of the supply chain. The interactions among these can be visualized as a series of gears, as in Figure 2.1. When adequately engineered, such a system can create breakthroughs, alternatively, when improperly engineered (or poorly maintained or unlubricated) such a system can lock up and become completely useless. Thus the theme of this chapter is the Gears of Change; we hope the world can achieve the smooth operation of these gears and not seize up the system, but we are not certain.

These Gears of Change will be used in a series of examples throughout this chapter and employed later in the book to offer predictions regarding the chances for success for many of the companies profiled in

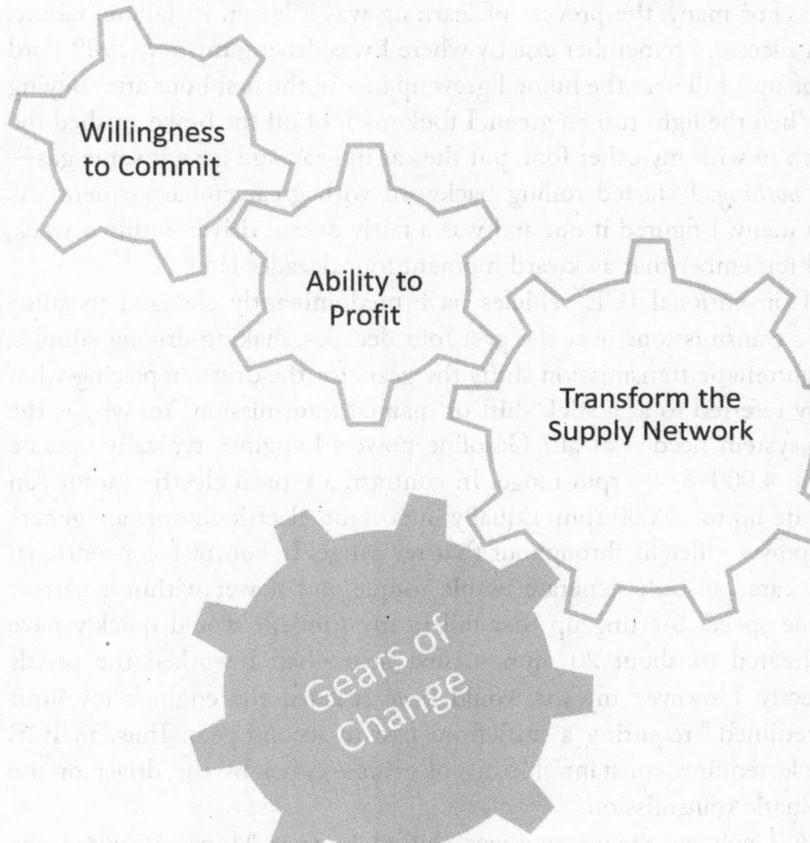

Figure 2.1. The Gears of Change

these pages. But first, it is useful to describe the intended pun regarding the Gears of Change.

Ironically, such a system for gears is unnecessary for EVs. Yet gears are a key component of ICE vehicles, one that is associated with the need for routine maintenance, including regular oil changes. One of the advantages of EVs is that they have far fewer moving parts.

Drivers of a certain age, and driving enthusiasts will remember driving manual cars where the driver had to push on a clutch and switch gears. For many, the process of learning was a lesson in failure, failure, then success. I remember exactly where I was driving my new 1989 Ford Probe up a hill near the home I grew up in—in the first hour after buying it. When the light turned green, I took my foot off the brake, pushed the clutch in with my other foot, put the car in gear, and gave it some gas—and *nothing*. I started rolling backward with great embarrassment. As with many, I figured it out and was a fairly decent driver within a week, yet I remember that awkward moment four decades later.

Conventional ICE vehicles have predominantly changed to automatic transmissions over the past four decades, making driving simpler. An automatic transmission shifts the gears for the driver, replacing what many referred to as a stick shift or manual transmission. Yet why is the gear system needed at all? Gasoline-powered engines typically operate in the 4,000–6,000 rpm range. In contrast, a typical electric motor can operate up to 20,000 rpm. Equally important, electric motors are generally power efficient throughout that rev range. In contrast, conventional ICE cars can only generate usable torque and power within a narrow engine speed. Starting up that hill in my Probe, I would quickly have accelerated to about 20 mph in first gear—had I worked the pedals correctly. However, my car would have reached the engine's rev limit or "redlined," requiring a shift from first to second gear. Thus, an ICE vehicle requires constant shifting of gears—either by the driver or the automatic transmission.

A Canadian steam engineer, Alfred Horner Munro invented the first automatic transmission using compressed air. The first production car to offer an actual automatic transmission was the 1948 Oldsmobile Hydra-Matic developed by Earl Thompson. Automatic transmissions did not become the predominant choice until sometime in the 1980s.[1] By comparison, electric cars accelerate to maximum speed in a single gear with little compromise. In addition, the lack of gears and the ability to rev the motor higher allows faster acceleration. A list of the fastest ten cars from 0 to 60 mph in 2023 included only five well-known names in ICE vehicles (Porsche, Audi, Ferrari, and Lamborghini) with price tags starting at $300,000. The remaining five spots in the top ten, including the fastest three cars, were all electric. Thus, one benefit of EVs is the ability to rapidly accelerate, which may be attractive to some consumers and boost their score of Willingness to Commit. As this chapter explores, getting the three elements in Figure 2.1 to work smoothly, as in a Lamborghini, is difficult, thus the transition to an EV future has many obstacles and opportunities.

The Nissan Leaf debuted in Japan and the United States in December 2010 becoming the first production EV to surpass sales of one hundred thousand units. The Leaf won the 2010 Green Car Vision Award, the 2011 European Car of the Year and the 2011 World Car of the Year awards. Until December 2019 the Leaf was the all-time top-selling plug-in electric car, only to be surpassed by Tesla Model 3 in early 2020. Global sales for this pathbreaker to date total 577,000. Beginning with a range of 73 miles, the second-generation Leaf introduced in October 2017 has a current range of 226 miles. This has been a successful introduction of a radically innovative new car. Yet how should we score its overall success? All of Nissan sold 3.9 million cars in 2021, so assuming that sales were relatively constant between 2010 and 2012, sales of the Leaf represent a relatively paltry 1.2 percent of Nissan's total sales.

In stark contrast, Tesla is all in on electrification, with 100 percent of its 3 million cars powered by electricity. Figure 2.2 shows how Gears of Change will be used to score predictions and outcomes.

I have assessed Nissan scores of 0.5 in the categories of *Willingness to Commit*, *Ability to Profit*, and *Transforming the Supply Network*. In this I use a 0–1 scale, with a score of 0.4 as the starting point at which the gear will turn. Higher scores indicate a better fitting, better lubricated gear, and thus a more successful outcome. Nissan clearly showed a willingness to commit in introducing the Leaf, some ability to profit, and reimagined its supply network, yet its success in each of these areas pales in comparison to Tesla. In my estimation, Musk and Tesla earn a perfect 1.0 score for willingness to commit and very high scores in the ability to profit and in transforming the supply network. In total, Nissan's aggregate score is calculated as the three scores multiplied together, resulting on a 0–1 scale with the Nissan Leaf being at 0.125. In contrast, Tesla's score is 0.81, based on its performance in all three areas and backed up by its high market valuation.

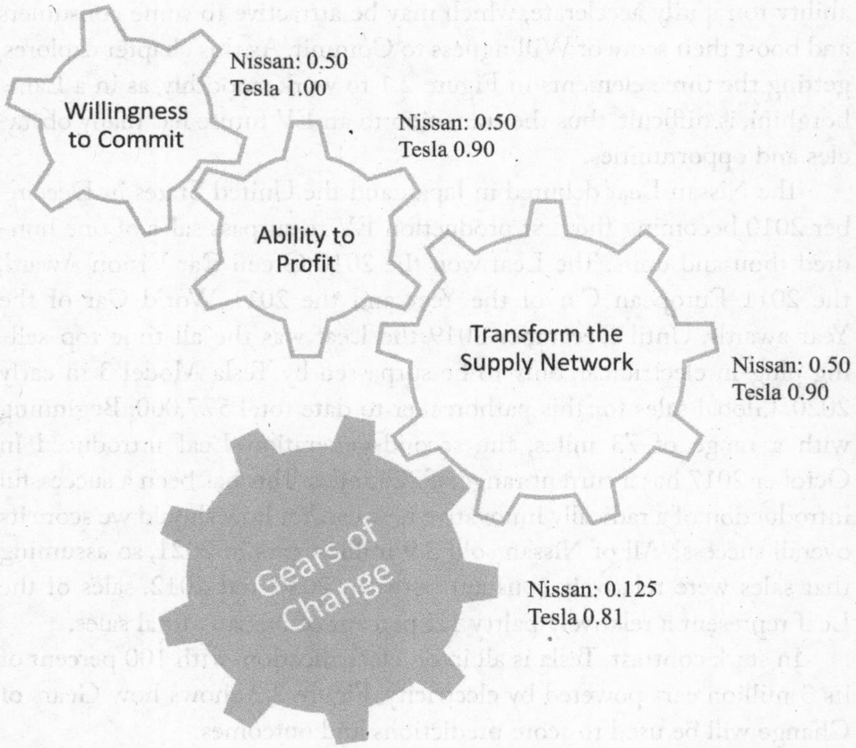

Figure 2.2. The Gears of Change: Comparing the Nissan Leaf and Tesla circa 2015

To be clear, these factors are rated subjectively by whoever is developing the ratings. In addition, the factors can be viewed from different perspectives. For example, willingness to commit looks different depending who or what is being assessed. Incorporated by Martin Eberhard and Marc Tarpenning on July 1, 2003, Tesla was intended by Eberhard to be a "car manufacturer that is also a technology company." Ian Wright became the company's third employee and the company raised $7.5 million in series A funding in early 2004. This is the point of arrival for Elon Musk as $6.5 million of the funding came from him following his windfall of $100 million from selling PayPal two years earlier. J. B. Straubel joined Tesla in May 2004 as chief technical officer. All four were fully committed to developing an electric car, yet the market was nonexistent and public policy was not supportive. Thus, willingness to commit differs greatly based on the eye of the evaluator. Similarly projections for ability to profit and transforming the supply network vary greatly. To truly transform the automotive industry, large groups of people, organizations, and governments must work toward some type of consensus—as emerged in the 1910s and 1920s as Ford and the moving assembly line produced and sold millions of Tin Lizzies.

GOVERNMENTAL JUMPSTARTS

The previous discussion examined willingness to commit from a corporate level in terms of the Tesla founders' vision, and implicitly in terms of the customers who were among the first adopters. While visionary leaders and advanced guard buyers can influence the trajectory of EV adoption, true societal change must be supported by governance. Laws, regulations and their enforcement have a huge impact on how corporations and private citizens spend their money. Thus, I focus on the three largest economies: the US at $21.4 trillion, the European Union at $19.2 trillion, and China at $14.1 trillion, all in terms of gross domestic product. Together these economies represent approximately 62 percent of world GDP. Equally importantly all three are seeking to foster and support substantial reductions in carbon emissions.

Aesop's fable "The Tortoise and the Hare" serves as an excellent analogy for China's emergence as the indisputable world leader in EVs. In the early 2000s China had become a powerhouse in manufacturing ICE cars having lured foreign manufacturers including GM and Toyota to bring manufacturing expertise and capacity to the country. Yet, like the Tortoise, China had minimal chance of catching the Hare(Japan) in a race for which both were vested and watching closely. According to He Hui, a senior policy analyst and China regional co-lead at the International Council on Clean Transportation, who opined on hybrids "[Japan] was already standing at the peak, so it failed to see why it needed to electrify [thinking] *I can already produce cars that are 40 percent more efficient than yours. It will take a long time for you to even catch up to me.*" The Tortoise [Chinese government] realized it couldn't win a race on the Hare's terms, thus in 2001 EV technology was introduced as a priority science research project in China's Five-Year Plan, the country's economic masterplan, a full two years before Martin Eberhard and Marc Tarpenning founded Tesla.

Here another auto aficionado brings personal leadership and charisma. Wan Gang was born in 1952 and studied experimental mechanics at Tongji University before traveling to Germany as part of his doctoral work at Clausthal University of Technology. Following the award of his PhD, Dr. Wan worked for Audi for a decade before making a visionary proposal in 2000 titled *Regarding Development of Automobile New Clean Energy as the Starting Line for Leap-Forward of China's Automobile Industry*. That proposal led to his return to China as a chief scientist for electric automobile projects for the Ministry of Science and Technology. In 2007, Dr. Wan was appointed minister of science and technology. He was also one of the first few dozen people to test drive a Tesla Roadster in 2008. He has widely been credited with propelling China into EVs. From 2009 to 2022 the government dispensed 200 billion RMB ($29 billion) into tax subsidies and tax breaks.

In addition to money flowing from the national spigot, several other Chinese initiatives were critical. First, local governments invested in public transit, buses, taxis, etcetera, which had a twofold effect of raising consumer awareness of EV technology and supporting the companies developing these technologies. The government also used a sweet tasting carrot, allowing people to automatically get a license plate for an EV in a country where for years they had been extremely rationed in large cities such as Beijing, with drivers having to pay thousands of dollars and often only being allowed to drive on alternate days of the week.[2] The result? In 2022 China was by far the largest market in the world for EVs, selling almost seven million compared to only eight hundred thousand in the United States. That year China exported almost as many EVs (679,000) as were sold in the United States. Furthermore, Chinese consumers have shown a huge willingness to commit with over 50 percent of respondents considering an EV for their next car in 2021, two times the global average. Very effective initiatives indeed, reflecting the hypothesized vision of Pulitzer Prize–winning author Thomas Friedman. In 2008, Friedman's closing chapter in *Hot, Flat and Crowded* was titled "China for a Day (but

Not Two)." In this summation Friedman postulated how a benevolent autocratic ruler might use totalitarian control to move China toward a greener planet.

Stepping away from communist China, the democracies of the United States and European Union allow citizens greater freedom and voice, and thus often face difficulty in developing consensus in willingness to commit. Indeed the EU has long been viewed as the world leader in environmental programs. Scholars Andrea Lenschow and Carina Sprungk argued in an academic article published in 2010 that "the myth of a *Green Europe* is successfully established and seems to be appealing to new generations. Yet, unlike in the case of foundational myths, storytellers have to make a double effort and show that Europe is both actually acting green and that it is 'destined' to carry a green mission."[3] The myth is a story that can motivate and bind disparate groups together, yet leadership in democracies is messy, as scholar Henning Deters argues "that the EU's unexpected dynamism and its recent decline are related to distinct institutional exits from the *joint-decision* trap that have opened and closed in different periods."[4] For our purposes, that decision trap is the ability to achieve near consensus on the need to act on transitioning to EVs.

The European Union agreed to a ban on sales of all new ICE vehicles by 2035 in October 2022. Is this part of the Green Myth, the power of magical thinking, or is it possible? Many are voicing their opinions loudly and emphatically. This agreement was hotly debated across member countries. In 2021 Germany, home to iconic brands including Mercedes-Benz, Volkswagen, Audi, Porsche, and BMW, the government set an ambitious target of having 15 million EVs on its roads by 2030; in addition, the country pledged to have 8 percent of electricity developed from renewable sources. Set against a total of only a million EVs on the roads at the time, this pledge had its share of skeptics. Ferdinand Dudenhoffer, a professor at the Center for Automotive Research in Duisburg stated, "The federal government's target will be missed by a wide margin. With the new funding guideline for electric cars, high electricity prices and rising battery costs, the market in Germany will collapse in the next few years." For established automotive manufacturers the pressure to maintain profitability is intense. At a recent New York International Auto Show, Dodge RAM president Mike Koval Jr. bluntly summarized the conflicting challenges:

> That's the elephant in the room for everybody. The cost of electrification is expensive, so for sure we need to make sure that we protect the profitability of our current in-market [internal combustion engine] business to help fund the transition to electrification.

In February 2023 Stellantis cited development costs related to the EV transition as a significant reason for idling a Jeep Cherokee plant in Illinois.[5] That plant is located in Belvidere, Illinois, about seventy-five miles northwest of Chicago in a county with about fifty thousand residents. The thirteen hundred Jeep employees are less than thrilled with this decision.

A prior round of subsidies ran into difficulties because German cars are sold on a global scale, thus the subsidies benefited other countries with less generous programs. Westward in France, the planned ban on ICE cars is facing stiff pushback and has been called "industrial destruction." Luc Chatel, president of the French car manufacturing trade union P.F.A., argues that a number of elements of the EV sector are not ready for the big transition, saying "France would need to have one million charging stations by 2030—940,000 more than it currently has."[6] This raises the joint-decision problem pointed out earlier. Individual and national concerns are a natural force that highly influences government policies. The European Union is a group of member states with heterogeneous membership that must forge some type of consensus, thus we see here concerns expressed in Germany and France, including natural desires to game the system and benefit one's own citizens. Interestingly, a feature of the Chinese government's approach allowed foreign companies to benefit from regulations. Tesla built its Gigafactory in Shanghai very quickly in 2019 due to favorable local policies. According to Tu Le, managing director of Sino Auto Insights, "To go from effectively a dirt field to job one in about a year is unprecedented. It points to the central government, and particularly the Shanghai government, breaking down any barriers or roadblocks to get Tesla to that point."[7] Today the Shanghai Gigafactory is by far Tesla's most productive, accounting for more than half of all cars produced in 2022. In comparison to China, both the European Union and the United States suffer from a substantial NIMBY, not-in-my-backyard, challenge.

In the United States, the passage of the Inflation Reduction Act in August 2022, created a wave of activity in the industry. One of the IRA act's primary features was the requirement that tax credits be applied only to vehicles produced in North America, not imported. This bit of nationalism is understandable from a political viewpoint, yet creates barriers to action. Likely to spur investments? Yes. Unclear exactly which companies qualify for those benefits? Also yes.

By October, automakers and suppliers had announced $15 billion in EV investments and retrofits of assembly plants, including Honda's massive $3.5 billion joint venture in Ohio with LG Energy Solution.[8] As I will discuss in chapter 6, Honda has long been a leading auto manufacturer worldwide and beginning in the late 1970s a major US employer. In fact Honda is a net exporter of cars from the United States. Among features of the IRA are the following[9]:

- Tax credits up to $7,500 per light-duty EV
- Tax credits on the sale of used EVs under $25,000
- Tax credits for commercial EVs of either $7,500 (under fourteen thousand pounds) and up to $40,000 for other vehicles.
- Tax credits of up to $100,000 for EV-charging equipment, with an important condition—the equipment must be in a low-income community or non-urban area.
- $3 billion to electrify the US Postal Service fleet, one of the largest fleets in the world with more than 235,000 vehicles

In terms of willingness to commit, the IRA is a very strong signal, which passed by the slimmest of margins in a 51–50 vote in the US Senate with VP Kamala Harris casting the tiebreaker. Then on April 12, 2023, the US Environmental Protection Agency announced an even more muscular set of goals/requirements, setting goals of two-thirds of passenger cars and a quarter of new heavy trucks sold in the United States being all electric by 2032. Bold, certainly. This is an executive action on the part of the Biden administration, one that can easily be reversed in the next national election in 2024. Michael S. Regan, administrator of the EPA said it was

> proposing the strongest-ever federal pollution technology standards for both cars and trucks. Together, today's actions will accelerate our ongoing transition to a clean vehicle future, tackle the climate crisis head on and improve air quality in poor communities all across the country. This is historic news.

Reactions from those in the industry were swift and combative. Consider John Bozzella, president of the Alliance for Automotive Innovation, which represents large US and foreign automakers. Bozzella challenged how the EPA could reasonably require companies to

> Exceed the carefully considered and data-driven goal announced by the administration in the executive order. Yes, America's transition to an electric and low-carbon transportation future is well underway—EV and battery manufacturing is ramping up across the country because automakers have self-financed billions to expand vehicle electrification. It's also true that EPA's proposed emissions plan is aggressive by any measure.
>
> Remember this: A lot has to go right for this massive, and unprecedented, change in our automotive market and industrial base to succeed.

As I write in the spring of 2023, there are governments, companies and individuals, betting trillions of dollars on this transformation. Things are going to change—but how quickly, successfully and smoothly? To continue the examination I next turn to Ability to Profit.

Ability to Profit

Of course, being willing to commit to something is different than being able to afford it. For companies there must be an ability to profit, which for consumers can be considered an ability or willingness to pay. To date, EVs have generally been far more expensive to buy, although these expenses may be mitigated by lower operating costs and tax incentives. EVs are typically much cheaper to operate on a daily basis for two primary reasons. First, electricity to power them generally is priced far below the equivalent cost of gasoline. Second, EVs have far fewer moving parts which correlates with lower maintenance costs and requirements. Design details differ from vehicle to vehicle, but a conventional ICE powertrain generally has over a thousand components. These include the axle, numerous gears and belts, and the oil to keep these parts running smoothly. In comparison, a BEV powertrain consists primarily of the battery pack, one or more electric motors that are more directly attached to the wheels and the controls for managing the power flow.[10] The relative difference in parts is roughly seven to one, and with fewer things that can break, maintenance costs and efforts should decrease proportionately.

Automakers long ago developed sophisticated strategies for luring customers. These include MSRP—manufacturer's suggested retail price (colloquially known as the sticker price)—dealer financing, and leasing. The existing system for selling and servicing automobiles has developed over a century and can be very tricky. Many, if not most, people dread buying a car for several reasons. First, it is a very high expenditure, and the common wisdom is that a car "loses 10 percent of its value the second you drive it off the lot." Not exactly tempting. Next, MSRP is what the automotive manufacturers suggest the car be sold for, yet dealers rarely sell cars for that amount. Dealers are provided an invoice price representing what they pay the manufacturer. At the same time, dealers do not have free reign. Automotive manufacturers also set a minimum price for which dealers can sell a car and assign vehicles using an allocation system. In short, consumers do not have anywhere near full-price transparency. Organizations, including Edmunds and Kelly Blue Book, offer consumers insights into vehicle pricing conditions. In February 2023, one-third of new vehicle sales in the United States were above the MSRP, with the average new car price as $45,296, almost 9 percent above MSRP. This situation reflects a lack of alignment between dealers and automobile manufacturers since any premium price goes solely into dealers' pockets.

At a time when the industry is seeking to leap from ICE to EV, this represents one more challenge and opportunity to overcome. As newcomers to the market, Tesla and VinFast can rethink how they sell cars and take the cars directly to the consumer. This creates many advantages while being limited by the need to build brand awareness, a fundamental hurdle for VinFast. By contrast, legacy manufacturers, including Honda and GM must navigate a path that works for both their goals and the goals of their dealer partners.

As if negotiating the base price of a car purchase isn't challenging enough, automotive manufacturers also have developed a sophisticated system of financing and leasing vehicles. In short, dealers offer consumers a longer-term deal to buy the car on an installment basis. A standard part of the purchase process is the negotiation around trading in an existing vehicle the potential purchaser already owns. If the purchase price of a new car is murky, the value of a used vehicle is downright muddy. In summary, car salespeople are generally about as popular as dentists—apologies to both salespeople and dentists.

The tidal wave of EVs offers the industry an opportunity to rebrand itself and reimagine the relationship with dealers. Indeed, there are challenges in selling a new type of vehicle to consumers with pricing uncertainty colliding with operating uncertainty. By operating uncertainty I refer to the fears that Henry Ford and the Model T Bible helped conquer in the early twentieth century. Consumers need to be educated on the benefits of EVs as well as how they work and should be driven and maintained. As companies jockey for position in this emerging market, evidence is emerging that profiles an industry that is rebranding itself.

Table 2.1 shows results from an Edmunds study of EV transactions during the first eleven months of 2022. This data examines three models that are seeking to dethrone Tesla—the Ford Mustang Mach-E, the Hyundai Ioniq, and the Kia EV6. These represent three of the four highest-selling EV models outside of Tesla. Each made significant sales gains during 2022 yet these models still trail Tesla by a large margin. Starting at the top of the table, the Edmunds data profiles the share of luxury trade-ins (i.e., high priced), these are the models consumers brought to the dealership to trade for a new vehicle. Dealers generally

Table 2.1. EV Transactions—First 11 Months of 2022

	Share of Luxury Trade-Ins for Model	**Share of Luxury Trade-Ins for Respective Brand**
Ford Mustang Mach-E	23%	6%
Hyundai Ioniq 5	17%	7%
Kia EV6	18%	6%
	Trade-In Conquest for Model	Trade-In Conquest for Respective Brand
Ford Mustang Mach-E	69%	42%
Hyundai Ioniq 5	81%	57%
Kia EV6	79%	60%
	Model Transaction Price	Brand Transaction Price
Ford Mustang Mach-E	$57,988	$55,609
Hyundai Ioniq 5	$54,643	$34,952
Kia EV6	$57,178	$34,651

consider luxury trade-ins as an indicator of the consumer's affluence and willingness to pay. A higher trade-in value makes it easier to get the consumer to sign on the dotted line and drive off with their new (hopefully high-priced) EV. The next section vividly illustrates how the sales process is viewed in the industry. A trade-in conquest is one in which the consumer moved from one brand to another—say from a Honda to a Ford.

Interestingly consumers have historically been very loyal to a single brand, with many being lifers. Country singer Toby Keith, a forever a Ford man, has been featured in Ford videos singing how he would rather walk ten miles than drive another brand truck and how he loves Ford Tough. A recent survey showed brand loyalty for twenty automotive manufacturers ranging from 38.3 percent of customers who would buy another Acura to the top three of Honda (58.7 percent), Toyota (60.3 percent), and Subaru (60.5 percent).[11] In short, the majority of consumers tend to repurchase the same brand, particularly if they have not had major problems with their existing car.

Most automotive manufacturers see a trade-in conquest rate of over 50 percent as a significant positive. Thus the data in the second section in Table 2.1 is powerful, indicating an average increase in conquest percentage from 53 to 76 percent in moving from the right-hand column (brand level) to the left-hand column (that specific car model). Conquest percentage? If it seems like the automobile industry is combative, that may be accurate. A "conquest" is when a consumer changes brands, say selling a Honda and buying a GM car. More important is the bottom line, in the last section. All three models showed a boost for the model transaction price (price for the new EV) over the brand transaction (the average price for all models sold by that brand).

Ford's increase was relatively small at 4 percent, while both Hyundai and Kia captured over 50 percent price increases (i.e., from $34,952 to $54,643 for the Ioniq 5). This illustrates one of the contradictions of the ability to pay: automotive manufacturers are generally seeking to sell more expensive vehicles to compensate for the immense costs of transitioning their supply networks to support EV manufacturing. At the same time, this puts pressure on consumers, particularly those of lesser means. Certainly it is the right of companies to sell what they can for the price they can, yet many worry about equity of access. Many public policy experts are keenly attuned to the dangers of leaving marginalized segments of the population out of transformations such as this.

Furthermore, there is the simple matter of the innovation adoption life cycle of the S curve. Typically presented as stages, including innovators, early adopters, late majority, and laggards, the status of EV adoption worldwide is in its nascence. Pricing of new vehicles will obviously have a huge influence on adoption rates.

To put some perspective on the impact of adoption rates on energy transformations, let's step back to the early 1800s. The primary means of heating homes in the United States was firewood. After all in a thickly forested, sparsely settled country, wood was cheap and abundant. Yet humans can be relentless; as cities grew rapidly, woodchoppers deforested areas around them. As early as 1744 Benjamin Franklin expressed concern in Philadelphia: "Wood, our common Fewel, which within these 100 years might be had at every Man's Door, must now be fetch'd 100 miles to some towns."[12]

As many today appreciate from a different perspective, a solution to the crisis sat below the very feet of early Americans in the form of anthracite coal, a dense, rocklike form. The appeal of coal lay in the fact that it was both wonderfully efficient—offering high British Thermal Units (BTUs) per pound—and that it didn't produce as much smoke as softer bituminous coal—or wood. Numerous entrepreneurs began ambitious projects to dig up coal and distribute it across the eastern and southern United States. At the same time few were willing to commit to this new fuel. For one thing it required a metal oven which was expensive and rare. Plus people hated metal stoves because they were enclosed, and you couldn't see or feel the comfort of the flames directly. In an 1843 short story titled "Fire Worship," Nathaniel Hawthorne argued for the status quo: "Social intercourse cannot long continue what it has been, now that we have subtracted from it so important . . . an element as firelight. While a man was true to the fireside, so long would he be true to country and law."

Yet the coal innovators were persistent. A combination of advertising, technological improvements and time led to widespread adoption. The 1860s represented when American households hit the tipping point and widely adopted coal. A family in 1831 would have spent the equivalent of $4.50 on coal for the winter while wood would have cost about $21, according to Sean Adams, a University of Florida history professor. Over time, as the coal industry grew, this cost advantage increased as the barriers for individual consumers reduced. By 1885 the new fuel had vanquished wood; Americans burned more coal than wood. According to Adams, "A lot of coal companies went bankrupt in the 19th century, it was an incredibly disorganized industry. These transitions take a long time and they are sloppy and they're intermittent."[13]

While the world watches today, the automotive industry is confronting similar, if more modern and fast-moving challenges. Just as the coal industry had to develop a new supply chain—from mining to transportation to retail to the stoves used to heat homes—automotive manufacturers in the current era also face many choices. Furthermore, one change does not prevent another. The first house I owned outside Chicago had a coal chute that had not been used in many decades, yet the house was also equipped with a modern furnace. Thus, I offer one more example of the inter-connection between willingness to commit and ability to profit before looping in the third gear, reimagining the supply chain.

The NFL's Super Bowl is annually one of the most watched live events in the world, broadcast in over one hundred countries with a viewership of over 120 million people. Advertisers pay huge premiums to showcase their product with a thirty-second commercials costing around $7 million. In most years, automobile manufacturers are the largest spenders, in 2022 alone spending almost $100 million. Yet, in 2023 the automotive industry drove off a cliff with only two companies, General Motors and Stellantis running commercials. Eric Haggstrom, director of business intelligence for Advertiser Perceptions said:

This has less to do with the Super Bowl itself and more to do with individual issues within the automotive industry. The auto industry has been battered by supply chain issues, inflation eating into consumer budgets, and rising interest rates that have made car payments dramatically more expensive.[14]

Here we see how the Gears of Change interconnect; 2022 marked a year where the automobile manufacturers went all in on the switch to EVs, hence, in one sense, earning a very high score on willingness to commit. Yet at the same time, these companies are challenged by part shortages and flow issues in the supply network for their existing base of ICE vehicles and demand was shrinking due to inflationary pressures. Something had to give. Having developed strategies for the coming decade that involve investments of billions per company, early 2023 marked a time when hedging bets was the prudent move, hence reduced Super Bowl advertising.

Transforming the Supply Network

The next focus for this chapter is on transforming the supply network. As explored earlier, the powertrain for EVs is substantially more straightforward, at least in terms of the number of parts, than that for ICE vehicles. There are many challenges in the development and production of batteries and other EV drivetrain components. Clearly, changes are coming for the supply network, and not solely in one vertical.

Chapter 8 examines the leading edges of power source exploration from electric to hydrogen. News coverage primarily focuses on battery powered, electric vehicles, yet there are many known limitations to this source of power. We will examine the challenges of transforming from a world based on gasoline to one powered by electricity. Challenges in transforming the power supply network include the original source of power—fossil-based or renewable, the storage and delivery of power, the battery and the end-of-life disposal or reuse of that battery. In chapter 9, we examine Monolith, a Nebraska-based company that has been refining via "methane pyrolysis," which takes natural gas, heats it to high temperature and splits it into two very valuable commodities—ammonia, the fundamental compound in most fertilizers, and carbon black, which comprises the majority of a tire. Monolith believes it can substantially reduce carbon emissions using its proprietary process and recently received backing from the US Department of Energy Loan Programs Office in the form of a conditionally approved loan for over $1 billion.[15]

Consider tires—as pointed out in chapter 1, the world produces and discards around two billion tires per year with vast environmental implications. Throughout the automotive industry, innovative individuals and companies seek means to manufacture tires more sustainably and capture used tires in a cleaner, more circular supply network. Chapter 9 examines Bolder Industries' quest to create a circular economy for tires and other efforts to green the wheels that move us.

While much of the world employs the term *supply chain* implying a smooth, one-directional flow of materials, I prefer the term *supply network* to reflect that materials and information move in many directions. In particular, as more emphasis is placed on circularity a linear supply chain clearly doesn't fit. As a place to start an examination of the need to reimagine supply networks, Figure 2.3 provides a high-level overview of GM's supply network. Automotive manufacturers typically classify their supply networks into tiers, with Tier 1 representing those suppliers that directly deliver parts, materials, or other services.

Tiers 2, 3, and beyond are the suppliers that deliver most directly to Tier 1 suppliers. GM's incredibly complex and diverse supply network comprises 538 companies reported to be direct suppliers.[16] I've shown the largest five suppliers by revenues earned in Figure 2.3. Lear Corp. is a huge corporation in its own right, with over $20 billion in revenue. Delivering finished sub-assemblies such as entire seats for a car or truck, Lear collects over $4 billion in revenue from GM annually, representing roughly 20 percent of its revenue. Normally, a supplier with a large (let's say 20 percent or more) portion of its revenue coming from a single customer is keenly attuned to that customer. In my core field of supply chain management, substantial research has been done on customer-supplier relationships and the impact of power differentials. The dominating theory is that manufacturers should develop a proactive partnership with their largest suppliers, seeking to create a Win-Win. Naturally, Lear executives seek to work in close partnership with GM as any changes in purchasing behavior can have outsize impact. In the current case, Lear is likely tracking changes in the automotive landscape very closely, yet an interior and electronics manufacturer is somewhat sheltered from many impending threats.

The second largest supplier to GM, American Axle and Mfg., represents a very different situation. With $2.3 billion and 40 percent of its

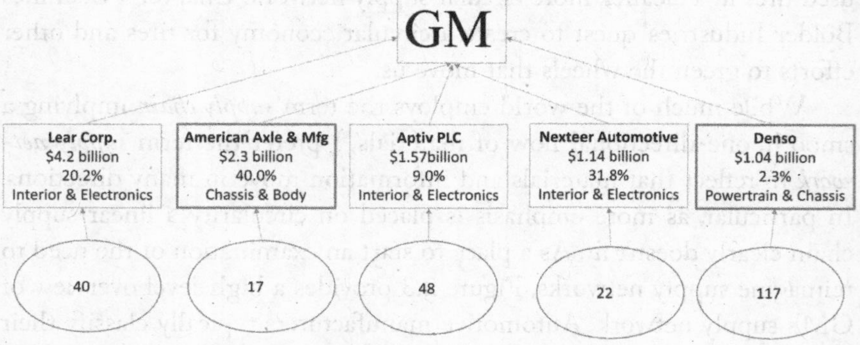

Figure 2.3. General Motors' Supply Network

revenue coming from GM, American Axle is already highly dependent on GM Adding in the fact that the company provides drive train parts that are changing rapidly and dramatically in the transition from ICE to EVs, and the company and its leaders clearly should feel their spider sense tingling. In Figure 2.3, American Axle and Denso are highlighted in red because these are two suppliers at the most risk as GM seeks to move fully toward EVs. That does not mean that all is lost, one way to look at every challenge is as an opportunity to grow, change, and develop new skills. At the same time, these companies have multi-billion-dollar investments in their existing manufacturing capabilities, thus their willingness to commit and ability to profit are very limited.

In total, the five Tier 1 suppliers shown in Figure 2.3 collectively earn $10 billion in revenue from just GM, and their other customers are likely also planning similar strategic changes. The ovals one level down represent the scale of the supply network, showing the total suppliers listed to each of the five Tier 1 suppliers. In total 234 suppliers are listed as direct to the five Tier 1s thus many thousands of jobs and billions of dollars in investments are on the line.

As a way to put the magnitude of planned change in context, consider my home state of Ohio. As the state with the second-largest automotive sector employment, Ohio employs more than 110,000 people. The state is also the second-largest manufacturer of transmissions and home of Honda's Anna engine plant, the largest in the world. As the industry transitions from ICE to EV, many of these jobs are at risk. Naturally, politicians and advocates are aware and concerned about this transformation. A recent report developed by JobsOhio in partnership with Benchmark Minerals Intelligence and the Natural Resources Defense Council (NRDC) examines the battery supply chain and advocates for Ohio playing a role as a "supply chain hub." The report examines GM's announcement of a planned 30 GWh cell facility to be co-sited with its Lordstown assembly plant. The JobsOhio estimate for job creation at this single plant estimates jobs in several stages of the supply chain ranging from primary jobs in the plant itself to tertiary—that is, jobs resulting from that plant—these can be such businesses as restaurants, financial providers, and grocery stores, as significant growth in one area leads to growth in other sectors.[17] The transformation to EVs is moving with such speed that the JobsOhio report issued in February, 2023 was already in print production and thus contains no reference to the earth-shattering (for Ohio) announcement in October 2022 that Honda will invest in a $3.5 billion battery plant in Fayette County in southwestern Ohio. In addition, Honda plans to spend another $700 million to retool three of its Ohio factories, including the Anna engine plant.[18]

Magna is another major supplier to GM, ranked by *Automotive News* as number 4 on the list of top 100 global suppliers. With revenues of $36.2 billion in 2021, it is a major player. Magna is not shown in Figure 2.3 because the databases don't always capture all customer/supplier relationships. The company builds battery enclosures for the GMC Hummer EV. A recent press release highlights the scope of change in the industry, with Magna stating that it planned to invest $700 million in two plants in Michigan and Ontario to produce these crucial components. The enclosures house and protect high-voltage batteries and other important EV components, and if they fail, they could cause fires. Note that the battery enclosures were not the root cause of fires that brought about GM's recall of all Chevy Bolt EVs. Magna claims its new manufacturing capacity will be able to "meet the individual needs" of its customers, with enclosures potentially made of steel, aluminum, or multi-material configurations, including lightweight composites.[19] Any technical or safety challenges with EVs are clearly damaging to customer willingness to commit to them, thus both GM and Magna have a lot riding on this venture.

This book focuses on four automotive manufacturers representative of very different stages in the lifecycle of the industry as well as dramatically diverse strategies. General Motors, the largest automaker in the world for seventy-seven years before losing its crown to Toyota in 2008, traces its roots to a holding company for Buick established in 1908 by William Durant. In a vivid illustration of creative destruction, Durant was the largest seller of horse-drawn vehicles as the twentieth century dawned, before leading GM as president from 1916 to 1920. In 1920, Alfred P. Sloane became the CEO of General Motors when Durant was removed due to financial imprudence. A marketing genius, Sloan initiated the establishment of annual model changes, which helped make prior models look "dated" and created a market for used cars. Sloan also developed the pricing model that has become nearly universal for car companies today, consisting of lower-priced, entry-level products and progressing to premium, luxury products. In increasing price, GM car brands in the mid-twentieth century included Chevrolet, Pontiac, Oldsmobile, Buick, and Cadillac. Of these American icons, both Pontiac and Oldsmobile have been discontinued.

As GM fights for market positioning as a legacy manufacturer, Tesla represents the other end of the continuum—the brash automotive unicorn with a stock market value higher than the next ten legacy companies combined. Starting a new car brand from nothing is a herculean task requiring enormous capital. In addition to Pontiac and Oldsmobile, there have been fourteen major brand failures since 1930 in the industry including Studebaker, Packard, Desoto, Plymouth, Mercury, and AMC. Each of these storied names was a profitable brand that sold substantial numbers of cars (at least one million each). Furthermore each of these brands had a lifespan of more than four decades. While the stock market imputes a huge value on Tesla prior success does not guarantee future success. In order to set the stage for more detailed examination of the newcomers, Tesla (chapter 3) and VinFast (chapter 5) versus the incumbents GM (chapter 6) and Honda (chapter 7), the following section tells the story of failed automotive brands.

Close Is Not Good Enough

A list of "brand-new" car manufacturers that have made an attempt at entering world markets since World War II is very short. Founded in 1945 as a partnership between industrialist Henry J. Kaiser and automobile executive Joseph W. Frazer, the Kaiser-Frazer company represented a bet that Kaiser's war time experience building ships for the US government would translate to success in automobiles. While Kaiser was new to the industry, Joseph Frazer had held positions with Packard, GM, Chrysler, and Willys-Overland. Ford, GM, and Chrysler were still producing models from before the war; hence, Kaiser-Frazer acquired a plant in Willow Run, Michigan, which was the largest building in the world. Built by the US government for Henry Ford to produce B-24 Liberator bombers, the plant offered production capacity that Kaizer-Frazier hoped to use to break into the market. Following some initial success, the market slowed in 1949 as the Big Three introduced new models, the company was renamed and became a holding company in 1953, exiting the industry after losing tens of millions of dollars.[20]

While Kaiser-Frazer was an ambitious and expensive failure, thus the name is unlikely to ring a bell for many, several other brands will likely resonate with readers. Many have heard of Ford's effort to launch the Edsel. Ford had been privately held since its founding, becoming a publicly traded firm on January 17, 1956. This meant the company was no longer entirely in the control of the Ford family, and Ford's new management thus had an opportunity for to expand its product range to match that of GM and Chrysler, which had a full range from entry level (Model A) to luxury (Model E). The car was named in honor of Edsel Ford, the founder's son. This did not sit well with Edsel's son Henry Ford II (a.k.a. Hank the Deuce—the founder's grandson), who was the CEO of Ford at the time. Despite Deuce's objections, Ford Motor developed and introduced the Edsel. While realtors often say location is everything, many in the automobile say timing is everything. Ford's timing was horrible, as the Edsel was introduced during a recession and was considered overhyped, unattractive, and produced with low quality. In sum, Ford quietly discontinued the brand in late 1959 following a loss of approximately $2.3 billion in 2021 dollars.[21] While dad Edsel's name went down in history as a byword for failure, his son Hank the Deuce served as Ford CEO from 1947 to 1979 and as chairman of the board of directors until 1980.

For a while, two car brands, Saturn and Hummer, launched by GM, were fairly successful. Saturn launched in 1985 and marketed itself as a "different kind of car company" to compete better with the rising tide of Japanese imports and transplants, including Toyota and Honda. The brand immediately established itself for "no haggle" pricing, winning millions of fans. Phased out of production as GM faced a deepening financial crisis, Saturn produced almost 3.7 million cars from 1992 to 2006. Saturn had been quite successful but eventually failed due to the inability of GM executives to adapt to new business ideas, leading *Forbes* to publish an article in 2010 titled "How GM Destroyed Its Saturn Success."[22] The Hummer began with American Motors Corp and was a civilian version of a truck designed to meet US military specifications issued in 1979. The first two Humvees produced in 1992 were bought by Arnold Schwarzenegger. In December 1999 AM General sold the brand name to GM but continued to manufacture the vehicles under contract for GM. AM General traces its roots through a history of the American auto industry, including links to the Kaiser-Frazer Company and AMC, both of which failed.

The Hummer rode a wave of success with the Terminator, with the H2 selling over one hundred fifty thousand units. Touted for being the "ultimate off-road vehicle," it developed a cult following. A 2006 review from *AutoBlog* gushed, "The H3 delivers serious off-road capability. But it also offers comfortable and quiet performance on the highway and is as capable at squeezing into tight parking spaces as it is picking its way over boulders on The Rubicon Trail. It's an impressive balance." At the same time, the H2 and its successors also had atrocious fuel efficiency with an EPA rating of 10–13 mpg in the city/country. The Hummer became the symbol of automotive excess that the Earth's climate could no longer afford.

On June 1, 2009, GM announced it was killing the brand as part of the larger bankruptcy announcement for the company as a whole. Much like the movie industry and Hollywood, the car industry and Detroit has a powerful drive to attempt to revitalize and employ a successful brand name from the past. In October 2020 GM unveiled its plans for an electric Hummer stating that "there is a lingering desire for the brand and what it represented."[23] At the same time, making such a beast run on electricity is no simple task. While some signs indicate the market is embracing the concept, in mid-summer 2022 the waiting list consisted of eighty thousand people likely to be frustrated as production was at a dozen trucks per day.[24]

The final memorable brand failure for the automotive industry is that of DeLorean Motor Company, which operated 1975 to 1982. A brilliant engineer, John DeLorean gained initial fame as the designer of the Pontiac GTO (Gran Turismo Omologato), a muscle car named after the Ferrari 250 GTO. In 1965 DeLorean was rewarded with a promotion to head of the Pontiac division of GM, at forty, the youngest man ever to lead a division. As division head, DeLorean led the design of the Pontiac Firebird as a competitor to Ford's "pony car," the Mustang. By the late 1960s, DeLorean was enjoying the fruits of his success, earning the equivalent of $3.5 million in present-day dollars. At a time when business executives in general and particularly those in the auto industry were conservative, button-downed types, DeLorean wore long sideburns and unbuttoned shirts. The best man at his wedding was another automotive icon, Ford president Lee Iacocca. In 1972 DeLorean was appointed to the position of vice president of car and truck production for all of GM (a position Jim DeLuca was to hold forty-two years later). Colleagues widely assumed he would rise to the presidency, yet this also rankled GM executives throughout the ranks.[25] When on April 2, 1973, he announced he was leaving the company many assumed he had been fired. DeLorean chose to position his exit differently, saying: "There's no forward response at General Motors to what the public wants today. A car should make people's eyes light up when they step into the showroom. Rebates are merely a way of convincing customers to buy bland cars they're not interested in."

Following his exit, he formed his own company, the DeLorean Motor Company, and designed a car initially called the DeLorean Safety Vehicle. The car's body was distinctive in using stainless steel and featuring two gull-wing doors opening vertically. A manufacturing plant was built in Dunmurry outside Belfast in Northern Ireland with financial incentives from the Northern Ireland Development Agency of around £100 million (which would be roughly $1 billion today). Renault was contracted to build the factory with over two thousand construction workers at its peak. In 1980 an American Express catalog featured an ad for a 24-karat gold-plated DeLorean. Aimed at the luxury market—by February 1982 roughly half of the seven thousand produced were unsold and DMC was $175 million in debt (over half a billion dollars today) and the brand was placed in receivership and discontinued. While the DeLorean gained a measure of notoriety as the time machine piloted by Marty in the film *Back to the Future*, it bankrupted its namesake. Nevertheless, a good brand is hard to kill—or to resist? Leading up to the 2022 Super Bowl, the "new" DMC released a teaser ad that an electric DeLorean was on its way to the market.[26]

Based on the history of automobile introductions over the past five to seven decades, the best uninformed or naive bet would have been on a failure for Tesla. Anyone making that bet would not be named Elon, who at this writing is once again the world's richest man. An investor foresighted enough to invest one hundred thousand dollars in Tesla on St. Patrick's Day 2014 would have held almost $2.5 million in stock in November 2021 and over $1.1 million on St. Patrick's Day 2023. This is the bet that VinFast, Vuong Vin, the team he has assembled is making with an estimated $9 billion investment to build its first factory in Haiphong, Vietnam. Beginning production of ICE vehicles in 2019 and transitioning to EVs in 2022, VinFast delivered its first forty-five sold vehicles to customers in California the first week of March 2023. A promising sign? Certainly. The company plans to build a factory in North Carolina to fully enter the US market and capture incentives offered by President Biden's Inflation Reduction Act. Can VinFast emulate Elon and Tesla? Its current financials, announced March 10, 2023, suggest the scope of the challenge—with 2022 revenues of approximately $630 million the company's net losses rose 55 percent year over year to almost $1.4 billion.[27] This is not a bet for the fainthearted!

The final automotive company this book examines is Honda, which exists somewhat in the middle ground between a legacy American company like GM and a new startup like Tesla. Honda was initially known as a Japanese manufacturer of motorcycles, becoming the largest worldwide in 1959. By the end of 2019, the company had produced an estimated four hundred million motorcycles and had also become the largest single producer of internal combustion engines with annual output of about 14 million. In 2013 Honda invested approximately 5.7 percent of its revenues into R&D (US $6.8 billion) and the same year it became the first Japanese automaker to export more cars (108,705) than it imports (88,357) to the United States.[28] At the same time, Honda was not always the global player it is today. Most of those cars are built at the Honda

Marysville assembly plant—less than an hour from my house. As the first Japanese manufacturer in America, Honda imported the large majority of parts, including the engine. The first Accord to come off the line was silver-gray and was given the license plate USA 001. In the final two months of 1982, Honda produced a total of 968 Accords; today, it can make over four hundred forty thousand per year.[29]

While Honda portrays this is a smooth, even journey destined for success, the reality suggests otherwise. In 2007 the assembly plant had produced nearly 9 million cars and represented nearly $4 billion in investment, including the largest focused engine plant in the world in Anna, Ohio. That engine plant today faces existential questions and threats to its very existence as Honda and the auto industry seek to transform themselves.

First produced in Japan in 1976 the Accord is currently in its tenth-generation design. At the Honda Research America building in Marysville, the company hangs the hood of an Accord for each new generation that its associates participate in the design of—a point of tremendous company pride. In total, Honda has sold almost 18 million Accords, including over 14 million in the United States and nearly 3 million in China. Following the first Accord to come off the line in 1982, Honda grew sales to take the crown as the highest-selling car in America in 1989, 1990, and 1991. Dethroned by the Ford Taurus in 1992, the crown went back and forth between Toyota (Camry) and Ford until Honda briefly reclaimed the bestseller title in 2001. After that, the best-selling US sedan has been the Toyota Camry for nineteen straight years.

From 1978 to 2017, General Motors took the top selling car title eight times (but never after 1987), Ford six times (but never after 1997). The dominance of legacy American manufacturers ended when Honda first took the title in 1988, with Honda and Toyota taking major market share from the Americans. Two things are important to highlight. First, Honda grew its sales following years of manufacturing the Accord in Japan. VinFast is seeking to copy and surpass Honda's amazing success. This is an audacious goal—the multibillion-dollar question is—what are the chances for VinFast's success?

One way to examine and vet VinFast's strategy is to examine a historical debate. In the 1990s, several prominent America researchers debated what was dubbed the "Honda Effect." While numerous papers have been written on the topic, the experts essentially divided themselves into two camps as argued and summarized by Richard Pascale. A consultant at McKinsey & Co. in the late 1970s alongside Tom Peters, author of *Good to Great*, Pascale wrote a book titled *The Art of Japanese Management* comparing Matsushita with the American company ITT with glowing reviews of Matsushita. This came at a time when there was great angst in the United States about the state of manufacturing and just as Honda was beginning to crest the wave in terms of selling motorcycles in the United States. In the spring of 1984, Pascale published an article in the *California Management Review* which ignited a debate that still raged over a decade later, one that is still relevant today. In short, to what extent was Honda's growth in the US market (first with motorcycles, then with cars and other forms of transportation) the outcome of an explicitly formulated and executed strategy versus the result of more of an intended but adaptable and learning strategy?

In 1975, the Boston Consulting Group presented the British government its final report *Strategy Alternatives for the British Motorcycle Industry*. The essence of this report argues that the loss of market share of British manufacturers in the United States from 49 percent to 9 percent resulted from a volume advantage. This means that as Honda produced more motorcycles, it moved down the learning curve and became more proficient at producing motorcycles for lower and lower costs. Pascale and other strategy experts argued that this interpretation was overly reductionist. In September 1982 Pascale convened a meeting of six Japanese executives who led Honda's entry into the US motorcycle market at the company's Tokyo headquarters. To briefly summarize his lengthy and informative writings, he argued that strategy was a more subjective, less linear process than BCG reported, writing:

> We (strategy experts) tend to impute coherence and purposive rationality to events when the opposite may be closer to the truth. How

an organization deals with miscalculation, mistakes and serendipitous events *outside its field of vision is often crucial to its success over time.*[30]

Strategy definition and execution is highly amorphous, relatively easy to recognize in hindsight yet very opaque in the present. At the same time, there are degrees of more right and less wrong. In a seminal article titled *The Core Competence of a Corporation*, professors Gary Hamel (London Business School) and C. K. Prahalad (University of Michigan) argued that the most successful companies focused on a few (no more than two or three) core competencies to achieve success. Honda's core competence was (and continues to be) the design and manufacture of internal combustion engines. According to Hamel and Prahalad: "Unlike Chrysler, Honda would never yield manufacturing responsibility for its engines—much less design of them."[31]

The core competence in ICE served Honda well for decades, but the times are changing rapidly. In April 2020 Honda and GM announced a joint partnership in which the two companies plan to codevelop EVs and their components, as well as hydrogen fuel cell powertrains. Under this agreement two EVs to be sold under the Honda nameplate will be powered by GM's Ultium batteries, while Honda will design the interior and exterior of the vehicles. Sales are planned for the 2024 model year with manufacturing to occur at GM's Detroit-Hamtramck plant, which GM Is investing $2.2 billion to retrofit for production of autonomous and electric vehicles.[32] In effect, the joint venture represents Honda's admission that it is behind in electrification technologies and needs to play catch-up. Presuming that the joint partnership produces vehicles that will do well in the US market, this leaves many outstanding questions and concerns for the ability to design, build and sell vehicles in Honda's home Japanese market.

As the remainder of this book unfolds, I am in search of insights on several fundamental questions keyed to this Honda Effect. First, what is the Tesla Effect, and how much of the EV market will it ultimately control? Second, where does General Motors fit in this transformative industry? Third, can Honda revise its core competence in internal combustion engines and make the directional shift to EVs? Finally, what are VinFast's chances of catching this wave and emulating the Honda Effect of three decades ago?

Chapter 3

The Birth and Evolution of Muskla

Origins

Before Elon, came M&M. Martin Eberhard earned a bachelor's degree in computer engineering from the University of Illinois in 1982 before beginning his career in the Silicon Valley. There he met Marc Tarpenning who also earned a bachelor's degree in computer engineering from UC Berkeley. Together they founded a company in 1997 named NuvoMedia, which developed and sold one of the first e-readers, Rocketbooks. This venture was moderately successful with Gemstar–TV Guide International acquiring it for $187 million in 2000. Meanwhile, Martin and Marc were set for life but not out of ideas, thus, the pair focused their entrepreneurial sights on building an electric vehicle.

Eberhard and Tarpenning founded Tesla Motors in July 2003 with a passion for cars and concern for dependence on imported oil and climate change. They served as CEO and CFO, respectively. The pair developed a set of three guiding principles:

- An electric car should not be a compromise. With the right technological choices, it is possible to build electric cars that are better cars than their competition.
- Battery technology is key to a successful electric car. Lithium-ion batteries are not only suitable for automotive use. These are game-changing innovations that bring a decent driving range into reality.

- If designed right, electric cars can appeal to even the most serious car enthusiast, as electric drive is capable of seriously outperforming internal combustion engines.

Ian Wright joined the M&M pair a few months later, with J. B. Straubel coming on board as chief technical officer in May 2004. Not until February 2004 did Elon Musk make his debut fresh off the $100 million bonanza from his PayPal sale proceeds two years prior. Musk provided almost 90 percent of the $7.5 million in Series A funds the group raised and became Chairman of the Board and Tesla's largest shareholder. This investment would eventually result in Musk becoming the richest human on the planet and J. B. Straubel becoming a billionaire. The other three founders did not benefit as magnificently from the financial upside of Tesla. Following a lawsuit initiated by Eberhard, a settlement was reached in 2009 that agreed that there were five founders of Tesla, in order of joining the company—Eberhard, Tarpenning, Wright, Straubel, Musk.[1]

The founders' strategy was to debut a premium sports car targeted at early adopters. This would allow premium pricing and an ability to profit (survive?) while building the company and transforming the supply network (really developing one) before a later pivot to mainstream vehicles and affordable compacts. Over the next few years, the company completed three rounds of financing bringing on board over $100 million in investment and such giants of the technology industry as Google cofounders Sergey Brin and Larry Page, as well as eBay president Jeff Skoll. Musk's deep connections within Silicon Valley and the venture capital industry undoubtedly were a foundational component of the nascent company's coming triumphs. At the same time, the drive, beliefs, and persistent passion of Eberhard, Tarpenning, Wright, and Straubel were elemental as well. Alas, there is a saying—"to the victor go the spoils"—and often, there can only be one victor.

THE ROADSTER: PROOF OF CONCEPT AND SOME BITTERNESS

The company's first car, the Roadster was intended to be an ultra-premium car priced at over $100,000when the concept vehicle debuted in 2006, roughly $150,000 today. By January 2005 Tesla had expanded to a couple dozen employees and was working on prototypes in a two-story industrial building in San Carlos, CA. The company finished a quarter-scale model that month, and a full-scale model in mid-April. By May 2006, the company was passing one hundred employees and believed it had a winning product. Tarpenning said at the time "You can feel it. It's a real car, and it's very exciting."[2] So exciting, that Arnold Schwarzenegger helped debut it with a photograph of J. B. Straubel driving with Schwarzenegger appearing in *Automotive News* on September 25, 2006.

One small problem. The difference between building a concept car by hand and mass producing one at any level of affordability is *huge*. Auto manufacturers routinely spend several million dollars hand-crafting a concept car that cannot be produced at scale. One primary reason is to tease customers with potential features and technology that might not reach the broad market for several years. One example was the Lincoln Navigator at the 2016 New York auto show, which boasted gullwing doors and steps that deployed when the door lifted open, almost inviting the passengers in. The gullwing doors were never added to a production model, but they did influence the trajectory of future Navigators. David Woodhouse, Design Director for Lincoln Motor Company, described it this way:

> Those really extreme ones (concept cars) are the hardest, but they would still have an influence on the market ahead. So you might not have seen them in the first couple of years after you first experienced them, but maybe the longer run, 10 years down the line, maybe they had a big influence on the trend of automotive design.[3]

In 2006, Tesla sought to show a concept car using electricity to a public addicted to internal combustion engines. The leap from a concept car to one produced at scale and sold to thousands of people was a huge one, far more significant than an established brand such as Lincoln putting the gullwing doors for the Navigator into high volume production. The reader may remember the many automobile brands described in chapter 2 that attempted to launch in America. Just as Arnold

Schwarzenegger was photographed driving the first Hummer—which had a great run—before being killed off, an early photo op for Tesla included J. B. Straubel riding in a prototype Roadster with Arnold. The terminator called the Roadster "hot." Having Arnold's endorsement is a huge marketing coup, but not nearly sufficient to successfully launch a new car brand.[4] Prior to Tesla, the last entirely new American brand still in play today is Chrysler, born in 1925.

A quick aside regarding Tesla's logo. Many people look at it and see a fashion-forward T, which is partly correct. In 2017, Musk offered a deeper level of insight when responding to a Twitter follower, noting that the logo is intended to represent a cross-section of an electric motor. Indeed, when the Tesla logo is repeated in a circle—it is an almost perfect representation of a cross-section of an electric motor. A great visual signal of Musk's willingness to commit to electrifying automobiles. In contrast, GM only modified its logo at the beginning of 2021 to emphasize electricity, redesigning its logo so that the gm is lowercase and the underline beneath the m makes the negative space inside look like a power plug. Deborah Wahl, GM's chief marketing officer noted it was only the fifth logo change in 113 years. At the same time, the industry stalwart was playing electrification catchup to Tesla.[5]

The Roadster that the company eventually brought to market drove exceptionally smoothly, but the road from concept car in 2006 to "success" was like driving a Tin Lizzie over an unpaved road—breakdowns happened—frequently. The first and second transmission/power management systems were failures, necessitating a third effort. Finding suppliers was like climbing Mt. Everest on one leg. One step was establishing a battery factory in Thailand, three hours south of Bangkok. The partner supplier used a building left primarily open due to extreme heat. While this worked well for the stoves and tires the company was making, it was an utter failure for the far more sensitive batteries Tesla required. Eventually, the partner paid over $70,000 to correct the conditions with temperature controls and drywall. Further, the Tesla expats spent hundreds of hours seeking to train the Thai workers on the intricacies of handling fragile electronics properly. This challenging episode was repeated with slight variations many times.

In August 2007 Eberhard, who had served as CEO was asked to step down by the board of directors and given the title "President of Technology" before ultimately departing in January 2008. Marc Tarpenning, who had served as VP of Electrical Engineering, left at the same time. Meanwhile, a website TheTruthAboutCars.com launched a "Tesla Death Watch." The company was burning through cash faster than an Olympic sprinter covering one hundred meters.

In December 2008, a fifth round of financing infused an additional $40 million, and in June 2009, Tesla was approved to receive $465 million in low-interest loans from the US Department of Energy. At this point, Elon had contributed $70 million of his own money, roughly half the payout he received from PayPal's sale to eBay. Willing to commit, indeed. Two of the cofounders followed their time at Tesla with additional efforts in the EV space. J. B. Straubel founded Redwood Materials, which seeks to produce battery materials for Li-Ion batteries via recycling. On June 21, 2022, Toyota and Redwood announced a collaboration to develop a closed-loop electrified vehicle battery ecosystem. At the time, Redwood processed 6 GWh of batteries per year, roughly enough to power one hundred sixty thousand cars, with a stated goal of growing to 100 GWh and 1 million vehicles per year by 2025.[6] In 2019 Eberhard founded Tiveni with the goal of making intelligent battery systems, which in turn was acquired by American Battery Solutions in September 2022.[7]

The classic product life cycle has four stages: birth, growth, maturity, and decline/death. The Roadster carried Tesla, Musk, and Straubel through birth. The Model S transitioned the company into the growth phase. The company produced and sold twenty-four hundred Roadsters, fewer than a week's production for a legacy manufacturer such as General Motors or Honda. Yet it was enough to keep the company alive for the Model S, which was produced and sold at a much higher rate. On New Year's Day 2015, the company was valued at $35 billion, having produced roughly sixty thousand cars—approximately $600,000 per car.

Maturing into a Powerhouse through Production Hell

In 2017 or 2018, Elon Musk coined the term *Production Hell* to describe what he and Tesla colleagues were struggling through. Journalist Neal

Boudette of the *New York Times* opened an article titled "For Tesla 'Production Hell' Looks Like the Reality of the Car Business" on April 3, 2018, with the following description:

> Tesla began producing electric cars at its plant in Fremont, Calif., six years ago, starting with small quantities and ramping up to about 100,000 vehicles last year. Now, as it tries to double or triple that number, the company and its chief executive, Elon Musk, are getting a lesson in how hard it is to mass-produce automobiles.
>
> On Tuesday, Tesla reported that it had managed to increase production of a crucial new model in the first quarter of 2018, although it remained well short of the company's already lowered target. At the same time, it encountered a new hitch—a drop in sales of its two established products, the Model S sedan and Model X sport-utility vehicle."[8]

Boudette tells a classic automotive story—getting all the Gears of Change to work effectively together is hard; a better description might be "next to impossible." Returning to the first line of the above article—yes, Tesla produced one hundred thousand cars in 2017, yet it only sold around fifty thousand. Locked gears.

Musk claimed to be sleeping at the Fremont plant, which it had acquired in May 2010 for $42 million. Stepping back in time, the Fremont plant operated by GM from 1962 to 1982 with atrocious quality and productivity. One of the leaders of the workers' own union, the UAW opined at the time that its workers were "considered the worst workforce in the automobile industry in the United States." Harsh stuff—and a mischaracterization. As demonstrated in *The Machine That Changed the World*, one of the primary principles of operational excellence is to examine the system, find flaws and defects to empower the worker to improve output. Clearly, GM was not doing this at the time.

Following the plant closure, GM and Toyota entered into a joint venture named the New United Motor Manufacturing (NUMMI) company. NUMMI was a transformative initiative that served as a learning laboratory for GM. It has been a case study for business schools around the world and produced over four hundred thousand vehicles in 2006. One might even say it saved GM, the subject of chapter 5. The journey was not easy. Tesla bought the Nummi Fremont plant in 2010 to serve as its first high-volume production plant. So this single plant, originally opened in 1962, has been a key component in GM's, Toyota's, and Tesla's journey. We will also see this phenomenon again when we discuss VinFast—automobile assets are capital intensive and specialized, but can be adapted and repurposed over time.

Returning to production hell in early 2018, Musk with his background in software did not appreciate the complexity of building a physical car with over five thousand parts from hundreds of suppliers. Musk envisioned a highly automated plant, yet that inevitably adds layers of complexity that are substantially different in the physical world than in the software field. According to Ron Harbour, an auto-manufacturing expert with Oliver Wyman at the time: "You have more new equipment to launch, there's more programming, more maintenance. More automation doesn't necessarily make it more efficient." Harbour added that the highest-volume plants he's seen often have more assembly workers and fewer robots. "It's a little counterintuitive, but that's how it is," he said.[9]

Another huge problem at the time? Matching demand for battery packs with a weak flow from Tesla's Gigafactory in Reno, Nevada. No power, no car. Chapter 8 will examine the intricacies of the battery supply network in more detail. For now, it is clear in hindsight that Tesla made it through production hell without suffering a death blow.

Not only did the company master or at least survive production hell, at the same time it managed supply network hell—or at least one of Dante's seven rings. The first Tesla Gigafactory, designed to produce batteries, was announced as a partnership with Panasonic on September 4, 2014. Founded by Kōnosuke Matsushita in 1918 as a lightbulb socket manufacturer, the Japanese battery giant initially planned a total investment of $1.6 billion to tool the first Gigafactory. On March 31, 2016, tens of thousands of people waited in *physical* lines in Australia, the United States, and Canada to place a deposit on a car *no one had seen* yet, the Model 3, which had been announced but not shown as a concept car. One week later, the company claimed it had over three hundred twenty-five thousand reservations, with about 5 percent of those including customers registering (and paying the $1,000 deposit) for two vehicles. Those customers included the author.

Panasonic announced a bond sale of nearly $4 billion in July 2016. Most of the funds were to be invested in the Gigafactory due to high demand. My willingness to commit was limited to a deposit—I did not order my Tesla until early March 2019, almost two full years after the Model 3 vehicles were delivered to mostly, but not entirely, happy customers.

In the interim, the Gigafactory began limited production of the Tesla Powerwall home energy storage device in early 2016. At the time, it used battery cells produced by other companies and assembled these into larger battery units. This demonstrates two elements of Tesla's strategy and execution over the years. First, the ability to partner with key suppliers to provide critical parts and learn and often build out its internal capabilities. Second, Tesla has not solely been a car company for many years. As Table 3.1 shows, the company has long manufactured power generation, transmission, and transmission products in addition to cars. This product breadth adds substantial complexity to the supply network, which is complemented by more significant revenue generation—if done "well." Of course, complexity often leads to material flow and quality problems.

Finally, Tesla also expanded internationally in an aggressive manner from 2020 to 2023, opening Gigafactories in China (2019) and Germany (2022) and announcing Gigafactory 6, which is under construction in Mexico. As shown in table 3.1, the Gigafactories typically mix the production of car components and charging/transmission equipment, with the benefit that Tesla has two separate lines of revenue, but with the additional challenge of managing diverse collections of parts and components.

Present: World Domination or One of Many Rivals?

At this point, I turn to an examination of Tesla and its current situation relative to competitors. On the summer solstice of 2023, Tesla looked pretty good by the financial numbers. The market capitalization was at $850 billion, a solid rebound from late December 2022 when the

Table 3.1. Tesla Production Facilities over Time

Facility	State	Country	Employment	Primary Use	Products
Tesla Fremont Factory	California	US	10,000	Transportation	Models—S, X, 3, Y
Gigafactory—Batteries	Storey County, Nevada	US	7,000	Transportation Power Storage	Batteries, Semi Powerwall, Powerpack, Megapack
Gigafactory New York	Buffalo, New York	US	1,500	Power Generation and Transmission	Solar Roof, Superchargers
Gigafactory Shanghai	Shanghai	China	15,000	Transportation	Models 3 and Y
Gigafactory Berlin-Brandenburg	Grünheide	Germany	10,000		Model Y (planned: batteries, Model 3)
Gigafactory Texas	Austin, Texas	US	12,000		Model Y, batteries (planned: Cybertruck, Model 3, Semi)
Gigafactory Mexico	Monterrey	Mexico			Planned: next-generation vehicle
		Total	55,500		

company had lost 75 percent of its value during the year. In June 2023, the company had posted $81.4 billion in revenue during the 2022 fiscal year and over $12 billion in profits. Experian data on EV registrations in the United States showed that Tesla registered a little over two hundred thousand cars during the first four months of 2023, or nearly 60 percent of the market. This compared to a little over 127,000 registrations in the year earlier period and 63 percent market share. At the same time, its competitors had roughly doubled total registrations from seventy-four thousand to one hundred forty thousand, and the total proportion of EVs in the market had grown to 7 percent.

So, what should we make of this data? Is Tesla on its way to world domination, or is it destined to be one of many players in a competitive market? To examine that quixotic question, the remainder of this chapter focuses on several key areas where Tesla seeks to maintain or develop a core competence over its rivals. This analysis will build on the Gears of Change.

Sales Channels: Willingness to Commit Confronts Willingness to Pay

In developing the Gears of Change, my choice of the category *Willingness to Commit* is a play on the term *willingness to pay* that consumer behavior and behavioral economics researchers have written thousands of articles and spent millions of hours studying. A search in a leading index of academic articles reveals almost forty thousand with the term in the title. It really needs no definition at its core, yet it is clearly context dependent. In the previous chapter, I profiled how the Nissan Leaf was the highest-selling EV for many years before Tesla pulled into the lead. The Leaf was a more useful, lower-priced car while, for at least the first decade, Teslas have been higher-end luxury vehicles for which consumers have paid a great deal more. These two concepts are tools to understand individual contexts and situations, but they are instrumental in drawing comparisons, as I will show. Keeping willingness to commit and willingness to pay in a healthy, gears-turning relationship is one of the key challenges for both Tesla and the EV industry over the coming decade.

An Ohio State colleague, Professor James Hill, teaches the popular MBA course "Matching Supply and Demand," which focuses on his expertise in supply chain. The course emphasizes the importance of delivering the correct supply of end products through developing and matching demand, highlighting the interaction between WTP and WTC.

A key element of Tesla's supply network and other upstarts, including Rivian and VinFast, is the absence of physical dealerships. One advantage is this allows the car manufacturers better control over pricing, sales, and delivery of the vehicles, as there is no middle person or intermediary. The academic literature labels this *dis-intermediation*, and much research has focused on the benefits and challenges. The fundamental early strength of Amazon was an inventory advantage based on not having physical stores in which as much as 40 percent of the inventory had to be returned to wholesalers and publishers because it couldn't be sold.[10] One of the things that Tesla has been able to do because it sells cars directly to consumers is adjust prices in a much more dynamic manner. A quick Google search reveals thousands of items on Tesla changing prices—both upward and downward—and many differences by market and region. In a seminal study, Erik Brynjolfsson of MIT and Michael D. Smith, now of Carnegie Mellon, coined the term *frictionless commerce*, finding that "Internet retailer's price adjustments over time are up to 100 times smaller than conventional retailers.'"[11] Tesla initially employed its dealership-free model to pocket a more significant percentage of the total revenue on products with a very healthy markup. In the past year, as competition heated up, Tesla is using dynamic pricing to match supply and demand. It is a tremendous tool to make these gears turn, yet it comes with the risk of increased consumer dissatisfaction, even anger.

An example of Tesla's use of price-tuning was in March 2019. They discounted the price of Model Xs in China while building Gigafactory Shanghai. That factory was situated in Shanghai for both labor-related advantages associated with manufacturing and market-related advantages tied to the burgeoning Chinese market for EVs. Ethen Hou, a thirty-two-year-old customer from Chengdu in western China, took delivery of a Model X sport-utility vehicle for which he paid 19 percent more than what Tesla changed the price to a few weeks after he purchased it. Mr. Hou referred to the $26,850 difference as "My contribution to Elon Musk's rocket-building project."

It takes little effort to find numerous stories on Google of frustrated customers complaining about the price changes. OK—but is this just the same as the customer/car dealer relationship with buying a new car and MSRP that was examined in chapter 2? Yes and no. The absence of dealerships means Tesla can pocket more of the proceeds of the sale and can also change prices more often and with less cost. In this situation, customers know to point directly at Tesla as a company (and usually Musk as the personification of that company) for any frustrations. Simultaneously, the absence of dealerships creates some disconnect and an inability to connect with customers physically. I will return to this, but first, back to matching supply with demand.

At the beginning of 2023, Tesla had a bloated inventory—many more cars were being produced than could be sold. By the end of Q1 2023, Tesla reported a 36 percent increase in US deliveries resulting from at least six different price cuts (and some price increases). During the same period, its stock market capitalization rebounded 131 percent. Fundamentally, Tesla was shaping the market to match its growing production capabilities. Still, the concern with this involves whether it is sacrificing its ability to demand premium margins to sell higher volumes of cars.

Across the industry, Kelley Blue Book estimated that the average transaction price for an EV was $55,488 in May 2023, a decrease of almost 15 percent from the nearly $65,000 average in the same period for 2022. While this is a substantial decrease, the EV average was still substantially higher than the industry average of $48,528 for all vehicles.[12] Greater competition among auto manufacturers and pressures to be under limits to receive incentives from the Inflation Reduction Act are substantial pressures for companies to reduce prices. At the same time, Tesla is in the driver's seat and employing its pricing power to grow its market share, at least for now.

At the same time as it flexes its muscle with dynamic pricing for its cars, Tesla has another considerable advantage in its charging network, in which it has invested billions of dollars and has provided a core competence for the company. The charging network also allows Tesla to flex its pricing muscles in another way. But first came an earth-shaking announcement from the White House on February 15, 2023. Namely, Tesla had agreed to open its proprietary charging network to competitors and their customers for the first time. This came through intense lobbying on the part of the Biden administration with Mitch Landrieu, White House infrastructure coordinator telling reporters on a phone call that the federal government is working to, "create a national network of chargers that will work for everyone, everywhere, no matter what type of car or state they're in."[13]

Is it promising for an equitable, lower carbon future in America? Certainly. Easy to achieve? Certainly not.

The Supercharger network that Tesla built and is now opening has been a critical foundation for its success. In addition to providing the possibility to charge quickly on road trips—potentially cross-country trips—early buyers also received free charging for life. For an average resident of Maine, this equates to an annual value of $1,343, yet only $627 for an average New Yorker. Either way, it is an excellent value, and for Tesla an easy-to-offer benefit (after the network is built out). Why do I say easy? Because this is the same principle as all-you-can-eat buffets—people buy with their eyes and spend with their stomachs, very few will eat to maximize their value! In Tesla's case, most people end up charging much of the time at home. After all, it takes less than thirty seconds to plug or unplug the charging and a sixty-mile range can easily be added overnight. Unlimited charging helped sell many cars, then in 2018 the company ended the perk for new car purchases in the future, claiming it was unsustainable.

Early in 2023, the company offered to "buy back" this perk by providing hundreds of thousands of owners a new, $5,000 discount if they traded in an existing Model S or Model X with unlimited charging access. Great marketing: Tesla gets to sell a new car while recapturing a "freebie" it had previously sold.[14] Another marketing move: On June 15, 2023, *Automotive News* reported that Tesla was offering three months of free, unlimited charging on its network for Model 3 sedans bought and taken possession of by the end of the quarter (June 30). Genius. First, this sounds like a great deal, but most consumers don't sit down and calculate out the exact value of something like this, rather they think "free? That's great!"

Further, consumers new to the EV experience likely don't realize any of the small and large difficulties of being tied (literally!) to the Tesla Supercharger network. Not that consumers can't charge at home, but it took the author over a year to change his ways and get to the point where over 95 percent of my charging is done at home in my garage. Also, to point out a concern with equitable access, what if a driver does not have easy access to a home charger? Returning to the offer of free charging, this is an excellent sales incentive that Tesla can quickly turn on or off and likely costs far less than what the consumer perceives as the value. In other words, it is a great tool to have in the company's toolbelt.

Transforming Supply Networks—Charging Standardization?

Starting in 2012, Tesla led the development of a two-part charging network in the United States, China, and Europe. Superchargers are located along highways and have four to fifteen charging stations, while destination chargers are in hotels, stadiums, stores, and restaurants. Tesla built all the Superchargers, while other companies built destination chargers, which charge slower but are simpler and less expensive to maintain. Note that "simpler and less expensive" to maintain is not the same thing as easy to maintain. A persistent challenge for drivers of EVs has been the unreliability of many destination chargers.

Before the agreement to open access to customers of other companies, the proprietary Tesla Supercharger network provided a significant advantage over competitors. In particular, the DC fast-charging network (so named because it operates on direct rather than alternating current and can charge a battery from 20 to 80 percent range in an hour or less) has provided Tesla with a huge advantage.

Introduced on September 24, 2012, with only six Supercharger stations, Tesla's network has grown to more than forty-five thousand Superchargers at over five thousand locations worldwide, including roughly two thousand each in North America and the Asia/Pacific region, with nearly one thousand in Europe. Each Supercharger stall is connected to an electric power supply of 72kW, 150kW, or 250kW. The Superchargers can refuel a Tesla to 80 percent charge within forty minutes. This network has been a foundational competence for Tesla's growing car sales, but it has not come cheaply. In 2017, UBS analyst Colin Langan estimated that Tesla would need to spend $8 billion to build a network that offered comparable convenience and access to the current network of gasoline stations in the United States for traditional ICE vehicles.[15] In a competing analysis, Loren McDonald argued that Langan made several fatally flawed assumptions. First, Langan ignores the heavy use of home charging, wherein 90 percent of early adopters' charging was done at home. Second, Langan's analysis did not incorporate either Tesla's network of destination chargers or public charging networks.[16]

So, where does the truth lie? Likely somewhere in the middle; indeed, Tesla has spent heavily building out its network, perhaps not $8 billion, but my estimate is at least $2 billion.

Here again, Tesla was willing to commit and felt poised to benefit through the ability to profit. Simply put, its network was developed because it was necessary to sell cars and was long held to be proprietary and exclusive to Tesla customers and drivers.

In February 2023, after "intense lobbying from the Biden administration," according to Shannon Osaka in the *Washington Post*, Tesla agreed to open its charging system to other users and began to call it the North American Charging Standard (NACS) It is widely seen to be substantially superior to the Combined Charging System (CCS) employed by most competitors in North America. Yet the multiple plug and charger types have presented a huge barrier to widespread customer adoption. The true extent of the tsunami of change became apparent when Mary Barra, CEO of the original automotive behemoth GM, joined Elon Musk on Twitter to announce a deal for GM and its customers to use the Tesla network. These bitter rivals fighting for ascendancy in the dynamic market sounded like friends.

Barra opened by saying, "When you think about what's happening in our industry and I've been in it for forty years, it is so exciting," with Musk countering, "Absolutely this is the most exciting time since Ford invented the moving assembly line." The pair of titans went on to talk about the "fantastic team" at Tesla, and Musk said what an "honor" it was to work with Barra and GM.[17]

This is a marriage of convenience, with many others joining. Ford began the tidal wave in late May, announcing it was partnering with Tesla, followed by GM on June 9, and Volvo, Rivian, and ChargePoint, a company providing charging services with CCS, joined on June 27. This follows a common pattern in the business world where a dominant standard "wins out" with the company that developed it profiting from selling or leasing the technology to competitors. An example from another industry involves the Amazon and Apple rivalry. When Amazon debuted the Kindle, it sold out of all inventory—in less than five hours. The Kindle's initial success propelled Amazon to become the leading seller of eBooks. Until Apple brought its bookstore out with the iPad in 2010, Amazon sold as much as 90 percent of all eBooks, and its market dominance allowed it to squeeze publishers on the prices at which they sold books wholesale. The iPad provided the same ability to read books on an electronic device—and much more. Thus, the debut of the iPad leveled the playing field by making it more of a duopoly than a monopoly. Now, the top publishers could play the two companies against each other. The key representatives of the leading publishers flew to Seattle within two weeks in 2010, demanding their ability to set prices for their books back. Amazon refused the first few, but soon realized it had to play ball and swiftly pivoted to develop an app on the iPad for downloading and reading books. In terms of core competence, Amazon did not see itself as a hardware company (at least at that point); thus, providing an app that would allow books to be bought from Mr. Bezos's company was just fine with Amazon leaders.[18] We are currently watching a similar act play out in EVs and charging. My supposition is that the world will end up with two, maybe three, at most four charging standards with NACS having already cemented itself in the lead spot.

Having said that Tesla is in the lead, it is important to note that the cement is not completely hardened yet. While Tesla is calling it the NACS, the S is the key part—a Standard is not really a standard until a group manages it and an eco-system of users has adopted it widely. Right now, that group is Tesla and Tesla owners. As reported by Jack Ewing in the *New York Times*, there will be a period of "corporate jockeying" over charging standards. Oleg Logvinov, the chair for North America for the Charging Interface Initiative (CharIN), which includes member companies such as Tesla, GM, and ChargePoint, among many others, says jockeying for position "creates confusion," and it is likely that customers "will probably wait until you can figure out which one wins."[19]

Tesla has successfully transformed the supply network for its own uses. NACS will become one of a handful of dominant standards for broader consumer and corporate use worldwide. This brings me to another aspect of Tesla's supply network where the company has developed a substantial advantage but where many limitations and questions remain—namely, with charger access.

A fundamental challenge and limitation for EVs is that whereas filling a car with gas can be accomplished in under five minutes, charging an EV is closer to an hour. Theoretically, it is possible to build a charging network where customers never run into lines (or at least long lines), but practically not so much. Researchers have long built operational processes around queuing theory, for which the bedrock foundation is Little's Law. Just like with Tesla Superchargers around Thanksgiving 2019, where drivers seeking to travel over the holiday encountered lines of well over thirty minutes to access a Supercharger, the difference between theory and reality is critical. The basic formulation of Little's Law (which predicts the number of customers a line can handle per hour and how long any lines will be) was published in a book by Philip M. Morse in 1958 as a *theory*. It wasn't until John D. C. Little, a professor at Case Institute of Technology, published a mathematical proof of the formulation in the journal *Operations Research* in 1961 that the concept began to be widely utilized and applied.[20] It was later named as a law and pretty much underpins most designs of corporate/customer interactions (i.e., lines) in the modern world.

Never fear, dear reader, I won't make you learn the math of Little's Law, but basic working knowledge helps develop a deeper understanding of the challenges of developing and EV network that will make long-distance drives fairly painless.

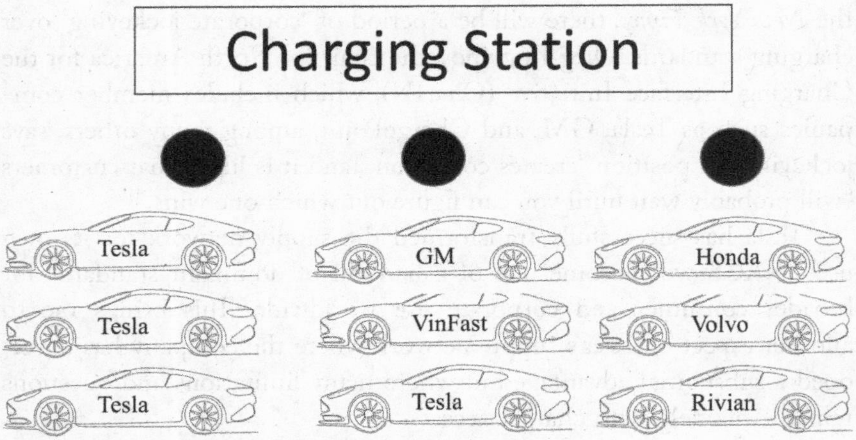

Figure 3.1. Lines at a Three Station Super Charger

Let's start with just the single line shown in Figure 3.1 (i.e., focus on the line to the left with all Teslas in it and ignore the middle and rightmost line). In academic terms, this is a M/M/1 queue, which means there is a single server (charging station) and a single line (in other words, customers line up nicely in one line, which is not always true in real life). Little's Law governs this situation. Let's assume the single line can process one car every forty minutes. This means that in a single day, that line can process thirty-six cars to charge for forty minutes each. Can we estimate the number of chargers and charging stations using this? Yes, to a point. First, the math gets complicated quickly when there is more than a single charge (as when we look at the whole of figure 3.1), and the *reality* departs from *theory* very quickly when human behavior is factored in.

To mention a single assumption of Little's Law, which rarely, if ever, holds true, customer arrivals must be evenly distributed. After all, a charger can't charge a car that is not there. In addition, there is a slight problem with the human beings queuing up—simply put, no one likes waiting in line! Two decades after Prof. Little published his proof of what became Little's Law, David Maister added a psychological interpretation to waiting lines in a case study published with Harvard Business School in 1984. I'll leave it to the reader to explore that further, but will offer Dr. Maister's quotation of a famous FedEx advertisement from the early

1980s: "Waiting is frustrating, demoralizing, agonizing, aggravating, annoying, time consuming and incredibly expensive."[21]

So how many chargers does the United States (the world) need to make the amount of waiting to charge acceptable to drivers? The answer—*a lot*! Consider when chargers are needed—estimates are that over 90 percent of charging occurs (or can occur) at home or work or plugged into a simple wall outlet. The problem is when a driver wants to travel further than the range of the car—then a charger with fast charging, reliability (i.e., works very close to 100 percent of the time), and availability is necessary.

So, let's work through a toy or hypothetical analysis of this. A 2021 news release from the American Automobile Association forecast that 50.9 million Americans would travel fifty or more miles from home for Thanksgiving.[22] Great, let's do the math. But wait. Some of these customers might travel just far enough to do it without recharging. My grandparents lived in Bexley, a suburb of Columbus. I grew up in Cincinnati, approximately 110 miles away. During the late 1970s and early 1980s, my parents would drive me (and my sister) north for Thanksgiving. If my parents had had my 2019 Tesla Model S, which has a range of 240 miles, they could have made the trip without recharging. Let's estimate that 25 million customers will travel far enough over the Thanksgiving holiday that they will need to recharge.

Plugging a forty-minute charge time into Little's Law shows that a single charger can charge thirty-six cars per hour, so we will need a little over 347,000 chargers nationwide if we assume that everyone will drive to their destinations on either the Tuesday or Wednesday prior to turkey day. Easy? Not so fast. That assumes the chargers are in the right locations and it assumes customers will arrive on a precise schedule. Imagine you are planning a trip from Dublin, Ohio, to LaGrange, Illinois for the holidays. You want to leave on Wednesday at 8 a.m. According to the Tesla trip estimator, the trip takes six hours and two minutes with a twenty-five-minute stop to charge in Fort Wayne, Indiana, assuming a full charge to start. Great, that's not bad. According to Google Maps, I can reach Fort Wayne in two hours and forty minutes, arriving at 9:40 a.m. (Central Time).

However, booking an appointment for charging is not that easy. Dr. Maister identified many uncertainties associated with human behavior. For example, people routinely skip appointments whether it is at a restaurant, doctor's office, or the gym. Thus, my earlier estimate of a third of a million chargers is too low. The real number of chargers needed is likely over six hundred thousand, and there are many other logistical challenges to consider. In short, here we see a major problem for everyone betting on EVs including Tesla—namely range and ability-to-charge anxiety greatly cuts into customers' willingness to commit. Thus while competitors adopting Tesla and NACS seem to offer an ability to profit, there are a many, many things that may lock these two Gears of Change up.

Margaret Atwood, author of *The Handmaid's Tale*, has been quoted as saying "You can only be jealous of someone who has something you think you ought to have yourself."[23]

Atwood's quote succinctly summarizes one of Tesla's conundrums, namely how will Tesla customers, who until at least 2023 generally paid a luxury car price for what most consider to be a luxury car, feel when owners of GM, VinFast, and Honda EVs show up in the same line that was previously exclusively theirs? My guess, as many in the news industry also guess—not great! This brings me back to Figure 3.1 with all three lines filled with a variety of different car models. Three things are true challenges for the future of EV charging in America and the world, as Little's Law and Maister's Principles predict:

1. There will be lines to charge EVs, particularly around holidays—and customers will get frustrated and angry (particularly Tesla customers who may feel they are special as early adopters).
2. There are many engineering challenges that remain to be solved, including how to charge vehicles made by scores of different manufacturers, each of which uses different powertrain technology (to be covered in the Power Sources chapter).
3. Battery and battery charger technology will improve over time, but it will take time. There will be struggles.

All three issues must be solved, either optimally, or at least mitigated. Otherwise, Tesla and the world will be stuck with a mixed situation in which early adopters are supplemented by the majority, but a substantial portion of laggards (or non-adopters) leaves us with a world in which ICE vehicles remain a substantial portion of the economy and leaves a suboptimal mix of fueling stations, some gasoline and some electric

Ability to Profit—Car or Computer?

As the final section of this chapter, I focus on the third gear of change. Tesla has shown an ability to profit, which raises the trillion-dollar question—what is Tesla worth in the future? A Google search of the company will turn up thousands of links where people are bullish or bearish. The company has been valued as highly as over $1.2 trillion (November 2021) and as I write this, its valuation is over $800 billion. For the sake of comparison, Tesla and Apple are in the same ballpark when their stock market valuation is divided by their revenues, with Tesla at 10 times value/revenue and Apple at 3.75. However, when one calculates their value divided by their profits (a.k.a. their price-to-earnings ratio) the difference is shocking. Apple's P/E ratio is 30, which is very healthy given the average for all Fortune 500 companies. Tesla's P/E ratio is over 200. This means that investors believe that Tesla will grow substantially in terms of revenue and profits in the coming years. This is not based on a bet that it will sell the majority of cars in the world, at least most financial experts don't believe this (I think).

As is often the case, the *Wall Street Journal* shed some bright light on this valuation when Stephen Wilmot wrote on June 21, 2023:

> There are a lot of theories for why Tesla's market value increased by about a quarter-trillion dollars during the past month. You have to go far down the list to find the thing investors usually pay most attention to: the foreseeable outlook for profit.

Wilmot's opinion is in the headline of the article: "The Problem with Tesla's AI Rally—Wall Street Shifts Focus from Price Cuts to Driverless Vehicles."

Full disclosure: I love my Tesla Model 3. It drives smoothly, accelerates like Bo Jackson from a standing start, and I have mastered (?) the charging process. At the same time, there are many things I would change or hope for in improvements. Both have to do with the subtitle for this section—it feels more like a computer than a car at times. Most of the things drivers have done with physical controls—turning windshield wipers on/off, changing gears, setting the air conditioning, and so on, are done via a huge touchscreen. To be fair, maybe I am getting old and curmudgeonly but the legacy manufacturers in Detroit, Japan, Germany, and other places had developed some pretty darn good technology for these functions over the past century. Other EVs I have driven feel much more intuitive and comfortable.

My second complaint involves the title of Wilmot's article, specifically Tesla's Full Self Driving (FSD) feature. To be clear, I bought it for a lot of money but rarely use it. Why? Frankly, it doesn't work to perfection, and it scares me—except when I am on a very long trip, and it is useful because I can use it to protect against my waning attention. I am far from the only person that worries that Muskla has overhyped the capabilities of its self-driving function.

In January 2023, the *New York Times* published an expose titled "Elon Musk's Appetite for Destruction" by Christopher Cox. The article sheds light on the history of FSD. It interviews and portrays both pro and anti-Musk fanatics and highlights the Reddit forum r/RealTesla, which showcases a plethora of FSD screw-ups and more generic complaints such as wrinkled leather seats, noisy cabins, and broken door handles. In my opinion, self-driving, whether developed and sold by Tesla or another company, is still a long way from being a safe reality. "Three weeks before Key [A rabid Pro-Musk fan of FSD] hit the police SUV [In his Tesla employing FSD], Musk wrote an email to Jim Riley, whose son Barrett died after his Tesla crashed while speeding. Musk sent Riley his condolences, and the grieving father wrote back to ask whether Tesla's software could be updated to allow an owner to set a maximum speed for the car, along with other restrictions on acceleration, access to the radio and the trunk and distance the car could drive from home. Musk, while sympathetic, replied: "*If there are a large number of settings, it will be too complex for most people to use. I want to make sure that we get this right. Most good for most number of people.*"[24]

The most good for the most people?

Who decides? The computer? This is an area humanity has not come to terms with yet, and likely will be grappling with for at least the next couple decades.

My personal forecast: I believe Tesla will grab a significant chunk of the overall market, but it will never approach General Motors' peak share of almost 50 percent of the American market in the 1960s. Why not? Simply put: too much competition and energy in this eco-system. Returning to the writing of Stephen Wilmot in the *Wall Street Journal*:

> Pinning hopes on Tesla robotaxis requires a double leap of faith because the technology is at an experimental stage and Tesla isn't among the most visible experimenters. While Chief Executive Elon Musk has talked about robotaxis, Alphabet's Waymo and GM's Cruise operate fleets of them, albeit in limited areas and conditions. GM and Alphabet moved forward with understandable caution.

The reader might notice the reference to GM and Cruise in the quote above. Not only is GM led by Mary Barra, who was named as one of *Time*'s most influential people in the world in 2014, but it is also partnered with Honda. Together these companies had revenues of nearly $300 billion in 2022 ($120 billion for Honda and $160 billion for GM). The two companies sold just over 10 million automobiles in 2022 and occupied the number 5 and 7 spots on the worldwide sales list. Granted the vast majority of their sales were traditional ICE vehicles, but counting them out of this transformation race is foolish.

Chapter 4

Forged in Flint

In *Star Wars: The Empire Strikes Back*, when Darth Vader famously tells Luke "It is your destiny," he may be unknowingly predicting the future for Jim DeLuca and Mary Barra, *née* Makela. Just like Luke ultimately triumphed over Vader with the power of the good side of the force, DeLuca and Barra have achieved great success in the global automotive industry and are vital figures in the revolution. They have worked together closely, learned, and led with passion and purpose. This chapter focuses on their intertwined careers beginning in 1979, the year before *Empire Strikes Back* came out, seeking to introduce these two giants before a deeper examination of the companies they have led or lead, namely (GM, VinFast, Ceer (for DeLuca), and GM (for Barra).

Jim Deluca was living in New Jersey in 1979, and his father had just been transferred (by General Motors) to Detroit. Jim was not happy leaving his friends in Bergen County, so when his dad suggested he apply to General Motors Institute, he thought he might be able to spend half of each year in Michigan and half in New Jersey at the Linden General Motors plant. His father challenged him, saying, "none of the guys I work with ever have their kids get into GMI," so young DeLuca took that as a challenge. Doubly so when he saw what his brother was facing in medical school—"I saw what he was facing and determined that I was probably thinking of medicine for the wrong reason—it wasn't for the pursuit of helping humanity; I thought I'd make a good living." Having impressed his father by getting admitted to GMI in Flint, Michigan, he started working at the Linden plant in July 1979 before moving to campus as part of the cooperative education program in the fall.

According to Jim, his father never gave him a lot of advice, but when he decided to enroll at GMI (later renamed Kettering University), his father, who spent his entire career at GM (starting his career as a mechanic in a dealership and rising up to be head of service in the western half of the United States for Oldsmobile) told him:

> "This is the only bit of advice I'll ever give you [about GM]. If you are a maintenance supervisor, production supervisor, sweeper supervisor, it doesn't matter. Just do your job better than anybody else ever could and you will never have a career discussion. You will never have to go to your boss and say, 'When am I going to get that next opportunity?' Because the opportunities will come, they always do when you deliver." I've never forgotten that, and I've told that story to many, many people that I've sat with over the years.[1]

At the same time, Mary Teresa Makela was still in high school in 1979, enrolling at GMI in 1980. Born the day before Christmas 1961, she grew up in Royal Oak, Michigan, and the automobile industry surrounded her; her father was a die maker for nearly four decades. She took her first co-op position, inspecting as many as sixty fender and hood panels per hour. Her earnings underwrote her education in electrical engineering, the same degree that Jim Deluca earned in 1984, while she graduated in 1985. Interestingly, Jim does not remember ever meeting her at GMI, saying:

> Mary's now husband, Tony, was in a fraternity out on the lake (and there are a lot of fraternities at GMI), and in five years, it's the one I never went to—that was a clique I never really interfaced with. I didn't meet Mary until my sister (also a GMI grad) introduced us at a GM conference; at the time Mary was the technical assistant to Harry Pearce. So I didn't meet Mary until long after we each graduated GMI.

Barra, for her part earned an MBA from Stanford while supported by a GM fellowship in 1990. Recognized early as a "high potential executive," she became Vice Chairman Harry Pearce's assistant in her early thirties.[2] Ironically, Harry and Mary share some history around product recalls, with Pearce profiled in a story in the *Los Angeles Times* in 1993. Growing up in North Dakota, Pearce had become fascinated by rockets as a teenager and got to know a lot about them. Eventually he moved on from his dream of becoming an astronaut to become an engineer and lawyer. But his early learning came in handy for GM when he forced NBC to admit that its *Dateline NBC* program had rigged a crash test using model-rocket engines to catalyze a fuel-tank fire in a GM truck.[3] Like her mentor Pearce, Barra would handle a quality problem masterfully during the first few months of her tenure as CEO.

Jumping forward to 2013, the year began with Barra, at the time senior vice president of Global Product Development, stating that GM would have half a million vehicles on the road with some form of electrification by 2017, focusing on plug-in technology. She said, "What started as a technology proof point has turned into a real-world starting point to push EV technology further and faster than we thought possible five years ago. The unique propulsion technology pioneered in the Volt—the same technology featured in the Cadillac ELR—will be a core piece of our electrification strategy going forward."[4]

By early October, Detroit and the press were speculating about the next leader of GM, with Barra considered along with three men as the next CEO. Coverage at the time recounted that "North American chief Mark Reuss and Global Product Development chief Mary Barra fought their way up GM's bureaucracy, survived the company's 2009 bankruptcy and together have 63 years at GM between them." The thinking at the time did not project current CEO Dan Akerson leaving until early 2015. Wrong! On December 1, the bombshell came that Barra would be the first female CEO of a US automaker, with pundits interpreting it as a positive move since the prior two CEOs came from other industries. Dave Cole, former chairman of the Center for Automotive Research at the University of Michigan said, "The key thing is the board and the CEO recognize it is a complex business and Barra knows it from the inside out; she's been around the block."[5]

Chapter 1 already covered the highlights of the ignition switch recall that provided a rude wake-up call for both President Barra and the company that had just named her CEO. I'd like to build on the previous story by connecting her career to Jim DeLuca's and discussing how they're guiding their companies toward a new era of electric vehicles. Let's begin with a brief timeline. December 10, 2013, Barra is named CEO of General Motors, the first woman to lead one of the American Big Three.

January 15, 2014, Barra gets the call from a senior colleague that GM cars had faulty ignition switches in late January. Small problem? Not. The defective ignition switches, traced back to cars made over a decade earlier, were eventually linked to at least 124 deaths and many other serious injuries. The worst part is that it later emerged that GM engineers had known about the defect—which disabled the power steering and airbags in certain situations while the car was operating—for more than a decade.[6] The ignition switch recall cost at least $4 billion, requiring compensation settlements for over four thousand customers, over $600 million of payments, and US federal fines of over $1 billion.

Returning to the timeline, Barra announced additional recalls on St. Patrick's Day for an additional 1.7 million cars and told employees in an internal video that "Something went very wrong in our processes in this instance and terrible things happened."

On the sixty-first day of her presidency, both GM and the president were facing unfortunate circumstances. Despite this, the president's courageous leadership, accountability for the issues, and compassionate meetings with affected families were already bringing about positive change for the future of GM. On April 2, Barra announced that attorney Anton R. Valukas would lead an internal investigation. The investigation ended in a report issued on June 5, 2014, which was day 141 of Barra's presidency. Within a month, she fired fifteen senior executives who were involved in the failure to understand how the car was built, according to Valukas's report.

This story is intended to highlight several points. First, manufacturing cars is difficult—extraordinarily difficult. Furthermore, mistakes in design or production have a very long life. The problems dated to more than a decade before the Comeback Queen's ascension. Fortunately for

GM, Mary Barra had and has a phenomenal leadership style that was well suited for the moment. Second, automobile production is an incredibly intricate and interwoven process, it cannot be accomplished without the skills, talent, and passion of thousands of people working together. Finally, as shown in the first three chapters, EV production is a bold new frontier in automobile production. Mistakes will be made, learning will occur, and the Gears of Change will turn.

Damage control is a form of leadership that is often necessary, but never fun or developmental. Barra also needed to assemble a team of leaders and change a culture that allowed the ignition switch recall to happen. One of her first moves was installing Jim DeLuca, announced on day eleven of her tenure. In our discussion, Jim described the selection process as:

> There were a couple of [high-level] placements at that time. I interviewed with three or four other very capable people for the position. But clearly, I was Mary's selection. I mean, I had worked with Mary in manufacturing at a time when Gary Cowger [former president of GM North America and mentor to both Mary and Jim] was leading the function, we both had leadership roles within manufacturing.
>
> And then when I went overseas [taking his family to Korea and China, with a one-year assignment in Thailand] and I was the head of quality for our international operations then, quality reported up through product engineering. And Mary then was leading product engineering. So I worked directly for Mary (Barra)—she and I had a very long and positive history. And so I was sort of [the] selection to go in to replace Tim Lee as the global head of manufacturing, manufacturing engineering, and labor relations.[7]

From Barra's perspective, she was confident she had hired a talented, experienced leader saying:

> Jim has a strong track record of aligning his team to deliver for the customer and drive results for the business. His global manufacturing and quality experience, along with his desire for innovative solutions, ensure

> we will continue our progress in producing the world's best vehicles, powertrains and stampings.[8]

Three decades following graduation from GMI, Jim DeLuca and Mary Barra were ready to reimagine the century-old General Motors. The pair would work hand in hand, with DeLuca reporting to Barra until his retirement in the summer of 2016. For the brainchild of Alfred Sloan, this meant getting the business on stable ground with both American and worldwide consumers following the twin threats of the Great Recession of 2008 and the faulty ignition switch recall of 2013. During this period Barra led her team to begin focusing on a longer-term strategy for electrifying its products which involves the extraordinarily challenging task of essentially running a very large manufacturing business successfully while also inventing a somewhat similar, yet unknown and dynamic new business. Following his retirement from GM, DeLuca took on first the challenge of building the first homegrown automobile company in the war-ravaged yet swiftly recovering/growing economy of Vietnam. After his four years with VinFast in Hanoi, he moved farther west to embark on the challenge of building an electric vehicle company in Saudi Arabia, which holds one-fifth of the world's conventional oil reserves.

Chapter 5

Vietnamese Visions

March 16, 2023, 4 p.m.—Student Parking Lot Vin University, Hanoi

Picture a tall, bald American walking through a parking lot conducting inventory on the vehicles. I am trying to get a picture of VinFast's market penetration from the center of Vinworld. As I walk up and down the rows, counting bicycles, motorcycles, and cars categorized as VinFast or other, a student pulls up on her motorbike carrying a handful of plastic bags filled with groceries. When I ask her if it is OK if I ask a few questions, she cheerfully asks me if I can guess where she is from. Hearing an accent that strikes me as Eastern European, I think Germany. No. Diana is from Ukraine, one of about a dozen students provided the opportunity to move to Hanoi, live at VinUniversity, and continue their studies using Zoom via links with their professors and classmates back in the Ukraine.

Pursuing a mechanical engineering degree and in what should be her final year of university, Diana has been granted free tuition and board to live at Vin University with her roommate, a first-year Vietnamese student. Diana has been tutoring English to Vietnamese children ages four to twelve to earn spending money. When asked about the war at home, Diana says her parents back in Kyiv are "just existing." She uses WhatsApp to talk with her parents and friends back in Ukraine, while devoting several hours each day to classes (Vietnam is over five hours ahead of Ukraine, so she needs to adapt to attend courses in a synchronous fashion). Diana says, "I am so happy I can live my life. I would like to stay here for now." Knowing she cannot get a visa for her chosen engineering career, she is considering applying for a visa to teach English.

Diana's story neatly encapsulates several aspects of Pham Nat Vuong's vision, as introduced in chapter 1. First, he is a brilliant man with an aggressive, ambitious agenda, much like Elon Musk circa 2008, already rich and established with visions of more. Here Vuong's vision includes more for his country and more for his many friends from Ukraine. Having made his initial fortune in Ukraine, he and his first company VinGroup acted quickly and generously in offering succor to refugees from the Russian invasion of Ukraine.

When Russia invaded in February 2022 a Vin University administrator, let's call her Erin Kotsi, was watching an interview with the president of an Ivy League university, her alma mater. That university was going to allow students to have extra time on their assignments and extend the deadline for paying tuition. Johnson thought, "Hmm, that is nice, but not much from one of the richest universities in the world." So in Hanoi Johnson approached the provost and president of Vin University and asked if there might be something the group could do to provide support for Ukrainians. A week later, a contact from VinGroup sent an email asking Johnson how many planes she needed. Following some fervent efforts to work with the Ukrainian Embassy in Vietnam, Johnson brought eleven female students to live and continue their studies in Hanoi. Chairman Vuong did not want any publicity or social media for this effort, yet at the same time one of the things that is not widely known is that there is a relatively sizable community of Vietnamese who emigrated to Ukraine in the 1980s and 1990s because they did not have opportunities to move to non-communist-bloc countries. Chairman Vuong feels an affinity to these people because they were his friends and colleagues when he was starting his noodle business, which served as a foundation for his future success. What does Vuong want today? To see Vietnam as a leader in EV manufacture and sales, saying, "Maybe not in five years, but in ten? We want to be at the top. Life is short, I cannot be slow."

Building a Supply Network to Manufacture EVs at Lightning Speed

Chapter 1 described how the assembly plant in Haiphong was built on land reclaimed from the sea in only twenty-one months. Chairman Vuong is an impatient man, and he expects everyone in his organization to dream boldly, work hard and fail fast, but learn from the failures. In describing the process of developing and positioning VinFast to be a leader in the EV industry, I rely on published news documents, which are relatively few compared to an established giant like GM, Tesla, or Honda. In addition, I had the opportunity to interview four people who work at or worked for VinFast. Jim DeLuca spent four years at VinFast, which required a special exemption from the Vietnamese government since work visas are only granted for three years. On leaving the company he joined Ceer, an EV startup in Saudia Arabia. Appointed as CEO on November 17, 2022, DeLuca joined a company with the mission of "creating a sustainable National Automotive Company that will ignite the KSA automotive ecosystem and generate a new source of valuable jobs for Saudi Arabian talent."

June 2016—In the Beginning There Was Jim

On June 10, 2016, General Motors announced that Jim DeLuca was retiring after a distinguished thirty-seven-year career with the company. Alicia Boler-Davis, a Detroit native, was appointed as his replacement, becoming the highest-ranking African American woman within the Big 3 Detroit automakers. A few months later, Jim received a call from a consultant with Boston Consulting Group, who had been advising a company in Vietnam and wanted to talk to him the next day. It was a Friday when Jim was relaxing after decades of hard work and thinking about what to do next. He replied that he had a family event and suggested Monday. But the consultant insisted on Sunday, which Jim thought was an odd day for a business call.

On Sunday he gets a call from Thuy Le who spent time at Lehman Brothers before it failed, then moved to VinGroup as CFO before becoming CEO. Thuy asked Jim how he would go about starting a car company. They discussed key strategic decisions such as licensing someone else's architecture, and how big an assembly plant to build and outfit with tooling. They talked for about two hours and at the end of the call she asked, "Would you like to join?"

DeLuca responded, "Join what?"

VinFast of course—which at this point only existed as a vague concept and in some strategic planning documents developed by Boston Consulting Group. After a little further prodding, DeLuca responded, "I just got back home with the family, we're happy to be home and I don't think I want to return to Asia." Madame Thuy asked if he would at least think about it. Out of politeness, Jim said sure and hung up to head home for Sunday dinner.

The next morning, he was out for a run in Clarkston, Michigan, listening to music on his headphones when his phone rings. It was Thuy. "Hey Jim, have you thought about it?"

DeLuca had to ask who it was. Then he responded, "No, we just hung up the phone." Thuy replied "Do me a favor. I know you've been to Vietnam with GM, but you haven't seen the things we are doing to advance the country—medical, schooling, property, shopping. Please. Come visit."

Two weeks later DeLuca boarded a plane and flew across the world. He spent a week looking at Vin Schools and VinMarts and listened to Chairman Vuong expound on his vision of creating an industry to elevate the country. Jim caught the bug, signed a three-year contract to be CEO, and set off to build a team.

DeLuca made his first two calls to bring coworkers on board. One employee reported, "In 2017 I got a call from Jim DeLuca, whom I knew from my GM days. He had been responsible for Asia-Pacific manufacturing while I was working on Asia-Pacific design." Deluca talked about the opportunity and having understood from his first couple of conversations with Madame Thuy that the Vietnamese business culture was impatient and all about speed. He said that they want to be selling cars within two years at a point when there is no factory, no team, no design, "There's no anything—right?" This employee signed up and joined the group, in no small part because of the allure of helping to design the first car and set the brand for a new company. The magnitude of challenges to overcome was incredible.

> We knew we wanted to do an SUV and a sedan. We knew they wanted to be in this E segment, which is basically a luxury class. But that's still like saying I have a dog, right? [What kind of dog do you want and where are you going to get it?]
>
> There are a lot of unknowns there. We had to go find an engineering firm. There wasn't a lot of interest [within VinFast] in building up the expertise internally. They wanted to just go and buy the talent and knowledge they needed—this is part of what BCG was telling them. You could just go buy an engineering and styling firm and certain things you don't have to have in house. You can just go and get them. It's turnkey excellence.[1]

The other employee was one of the first hires for VinFast. He joined even before Jim's contract was fully executed after receiving a call from DeLuca. A component of his valuable experience was the two years he spent in the early 1990s at the Nummi production plant partnership between Toyota and GM. In this venture GM learned much about the Toyota production system that helped it adapt to the auto environment of the 1990s and early twenty-first century. Then Tesla bought the plant to make it the first high-volume production facility for producing EVs.

By the end of 2017, DeLuca had assembled a team of a couple hundred experts drawn from across the auto industry, primarily from Western countries. Many of these experts worked remotely and traveled to Vietnam for periodic in-person events. According to the first employee, "I was living in Detroit and just going on business trips occasionally, sometimes Vietnam, sometimes to Italy, where the styling work was happening, sometimes to Austria, where Magnus Steyre was doing engineering."

At the same time that the core VinFast team was being built out, two other fundamental problems had to be solved. First, what would the product be—that is, what kind of car was VinFast going to build and sell? Second, how would the new company build its cars? They didn't have a factory or the beginnings of one.

Focusing on the product first, the Americans mulled over building a luxury vehicle employing as much outsourced development as possible. The first order of business was to find an architecture (essentially the chassis and power train) for an existing car that VinFast could obtain the rights to and put its own interior and exterior components on. According to DeLuca:

> So we started shopping around and we met BMW at the Frankfurt Auto Show and we started to talk to them. I was very impressed with BMW because they are a big company, they're a very bureaucratic company. Yet they were able to talk to us in a way that demonstrated that they'd be able to move in concert with us.
>
> They knew that we were going to move fast, and they were fully supportive. So we pretty quickly ink[ed] a deal with BMW. We licensed three different things. We licensed the X5, so we had the SUV, we licensed the five series, so we had the sedan. And then we licensed their six-cylinder engine. So we had the power train.

To hear DeLuca tell it, it sounds like they went to Nordstrom to buy a new suit. Negotiating and signing such deals is a very complex enterprise. In some ways, it's like dealing with the mafia: many soldiers, few leaders. "This is one of the values VinFast got in Jim DeLuca—everybody knows him. He's legendary; if he calls Robert Bosch (A German automotive parts company with revenues of 78 billion euros) or Faurecia (A French firm that is the ninth-largest international automotive parts manufacturer), or BMW, they take his calls. They then say, '*Is this really going to happen Jim?*' He is one of the few people in the world with the credibility to get suppliers to trust and bid on these jobs," says one employee.

To create an automobile company where nothing existed, VinFast followed parallel paths. First, it talked to and negotiated with several established manufacturing and design companies to subcontract components to design its own vehicles. Second, it secured licensing rights with Jim's former employer GM. Two announcements came in June 2018, stating that Vin Group subsidiary VinFast had contracted exclusive rights to distribute Chevrolet's within Vietnam. Further, VinFast would assume ownership of GM's Hanoi factory to produce small cars. DeLuca stated, "We are excited about the opportunities presented by this announcement." The plant produced 7,600 vehicles in 2017 but has the capacity for 30,000. This is significant news in a country where motorcycles are still the primary form of transportation. However, for GM, it may be one of many relatively small deals it makes annually.

John Reed, correspondent for the *Financial Times* wrote:

> For Vingroup, founded and chaired by Pham Nhat Vuong, Vietnam's richest man, the deal with GM will mark a deepening commitment by the sprawling company to the promising but risky business of making cars. Vingroup styles itself as a "cradle-to-grave" provider of goods and services to Vietnamese consumers with businesses in retail, real estate, education and other areas.[2]

In parallel with the deal DeLuca signed with GM and his former boss Mary Barra, he also led the efforts to acquire an architecture, signing a deal with BMW to acquire rights to three older BMW models. Madame Thuy, DeLuca, and people on the VinFast team believed this deal was great since it allowed VinFast a relatively fast ramp to producing automobiles in Vietnam. The plan was to use the GM plant in Hanoi, construct a larger and more modern facility in Haiphong (roughly eighty miles east of Hanoi), capture a significant share of Vietnam's automotive market, and later expand to the United States and Europe. This a very reasonable plan, very similar to the strategy that Honda and Toyota followed in first establishing their brands as preeminent in Japan and then expanding to become world players by moving overseas. Hyundai and Kia followed similar strategies in the 1990s and early 2000s. The audaciousness of the plan lay in the speed with which they sought to execute it. Honda was the largest manufacturer of motorcycles in the world in 1959. From this solid base, it took the company twenty-three years before it rolled the first Accord off the assembly line in Ohio. By comparison VinFast is seeking to go from nothing to full-scale assembly plant in Vietnam and another plant in North Carolina in under a decade. Monumentally audacious. Furthermore, as examined in chapter 1, strategy is both intended and realized, and events rarely unfold in a scripted, predictable manner. Thus, adaptation is critical.

The deal with BMW offered VinFast an excellent opportunity. BMW produces beautiful, luxurious cars, thus developing its own interior and exterior on a BMW chassis allowed employees, VinFast, and the Italian design company Pinnafarina that it contracted to design a beautiful ICE car which has generally been very well received in Vietnam. Yet it came with other challenges. When asked if BMW had been willing to license the technology because it was, literally, an out-of-date model, DeLuca responded:

> If you asked me the question today, the answer is different than what we did in 2017. In 2017, you're exactly right. They were too protective of the launch product. So what we licensed was the prior product. Little did we know that was gonna be a huge problem. Why is that? All of their suppliers were moving then to the new product, and many of their components now were changing. And so when we went to the suppliers and we'd say, "Hey, we need this unit from the old X5." They'd say, "Oh man, Dude, we already retooled."[3]

Retooled in the auto industry is shorthand for, "Yes, we can produce that part for you, but it is going to cost. Millions of dollars." To make a long story shorter, Madame Thuy, DeLuca, and the VinFast team forged ahead through all of the associated challenges, building the assembly plant in Haiphong, one of the world's largest, most vertically integrated automobile production facilities. Comprising 860 acres and almost 9 million square feet, the facility employed experts from over forty countries when I toured in March 2023. Employing over twelve thousand people in sixteen different shops, the VinFast facility in Haiphong represents an investment of nearly $9 billion. VinGroup, Vuong, and partners are investing very heavily in this enterprise. During the tour, Deputy CEO Michael Johnson pointed out that this was the first instance of heavy-duty discrete manufacturing in Vietnam, making it a strong economic engine for the country's aspirations. By 2021 VinFast was a reasonably successful automobile manufacturing company with a state-of-the-art plant and was selling cars based on the BMW architecture as supported by Magna in Germany for engineering and Pinnafarina in Italy for design. At the same time, Chairman Vuong did not become the richest man in Vietnam without taking risks and pushing for more.

Jim DeLuca and everyone that works for VinFast learns to agree with Mr. Vuong, get out of his way, or pack your bags to leave Vietnam. According to DeLuca, a story illustrative of the national fervor in Vietnam and support for everything VinGroup:

> I'm in the Metropole in Hanoi. I'm going up in an elevator and there's a young hotel manager. We were there for a conference. I had a VinFast badge on. He said, "Do you work for VinFast?" I said "Yeah." He said, "I just want to thank you for what Vin group has done for Vietnam." He said, "I'm living a life. I never thought I'd be able to live and thrive in Vietnam and it's because of everything Vin group has done. I live in a Vin home. My kids go to Vin school. I shop at Vinmart. I work here at the Metropole. But all these opportunities have been made available to us by the vision of the chairman."
>
> I thought, "Wow, here is an unsolicited accolade for Vin Group."

The culture at VinFast is aggressive, focused on speed to market, and extremely top-down. This type of culture can do amazing things—such as get a huge manufacturing complex in Haiphong built on land reclaimed from the sea in only twenty-one months. At the same time, a fail-fast, learn-fast culture and strategy also can lead to gigantic problems in the expensive, complex world of automobiles. In April 2022 *Motor Trend* gushed that VinFast was making car manufacturing "look easy,"[4] and in March 2022, when President Biden tweeted about VinFast's plans to build and open an assembly plant in North Carolina he said it was the "latest example of my economic strategy at work." Maybe. . . . But not yet!

As 2023 unfolded, it seemed as if a string of events involving VinFast was written for a show titled "Trying to Come to America." The strategy from the beginning was for 40 percent of the company's sales to come from the United States and 40 percent from Europe. There is not a sufficiently developed economy in Vietnam to support a company with a facility designed to produce a quarter million cars annually. Thus, to operate at scale, VinFast had to export cars to the United States and Europe, much like Honda and Toyota starting in the 1970s. Only VinFast and Vuong wanted to get there faster.

It was Christmas in November 2022 when the company announced it was shipping 999 of its VF 8 cars produced in Haiphong to the United States to begin delivery to a purported sixty-five thousand customers that had booked orders.[5] Shipments to Europe were projected to follow six months later. On December 5, the company excitedly announced the opening of stores in Cologne Germany and Paris. Yet, not until March 2023 were the first cars delivered to customers, with forty-five of its VF 8 City Edition all-electric SUVs delivered to US customers at its nine stores across California. Gareth Dunsmore, VinFast deputy CEO for global sales and marketing said: "The event drew tremendous attention from our US consumers. It's extremely gratifying in fulfilling our delivery promise and to witness the excitement our customers have for the VinFast brand. This is a great moment for all of us and more motivation to continuously strive to exceed the expectations of our valued customers."[6]

But the wider public did not believe Mr. Dunsmore's rhetoric. The reviews were horrible. *Car and Driver*'s take was that the "2023 VinFast VF8 Proves Building Cars Is Hard," while *Motor Trend's* headline was "Return to Sender." Ouch. A YouTube influencer, Donut, with 8 million followers, posted a review in July 2023 titled, "We Drove the Worst Reviewed Car in America."[7] The pair of reviewers carefully work through a checklist of problems that VinFast customers have reported, concluding that some were overblown and some were accurate. Two of the biggest concerns? A lag in acceleration on the highway, with the note that the VF8 might be the worst car to escape a crime scene since it took a full second or more once the accelerator (not the gas pedal) was pressed for the car to speed up. The biggest concern? Likely the one reported by Tom Peng, who experienced a complete shutdown in the VF 8 he was driving—the car thought it was driving when it was not. In conclusion, Donut leaves viewers with a negative review of the VF 8.

Early reactions to VinFast's American entry leave much to be desired, but are these substantially different from what Toyota and Honda faced in the 1970s and 1980s? Or Kia and Hyundai a decade later? Certainly the internet has enabled an environment where lovers and haters can broadcast their opinions much more widely than earlier efforts. So how should VinFast respond?

The company is working to address quality issues using software updates. One key advantage/challenge of EVs pioneered by Tesla is the ability to modify software to address many issues, as long as the problem is not mechanical. VinFast has issued several communications to customers and the public regarding efforts to improve vehicle performance. In May 2023 VinFast issued a voluntary safety recall of the VF 8 City Edition on which the dashboard screen goes blank while driving or stationary. "VinFast is not aware of any field reports of incidents. The company is issuing this recall out of an abundance of caution." The recall came after the US National Highway Traffic Safety Administration said 999 VF 8 vehicles suffered a software error in the dashboard display that prevented critical safety information from being shown and "may increase the risk of a crash." The NHTSA went on to estimate that the company still possessed seven hundred of the original shipment of cars, meaning it had only sold about three hundred cars in the United States in four months.[8] That recall followed another in February of almost 2,800 units over an issue with the front brake on some cars. Not ideal, but not necessarily a death sentence either, as the reader may recall with GM's ignition switch debacle.

Vietnam and VinFast can look roughly 2,700 kilometers northeast for inspiration and an illustration of the Lazarus effect. Kia Motors was founded in 1944 as a manufacturer of steel tubing and bicycle parts, later expanding to produce Honda-licensed motorcycles in 1957 and Mazda-licensed trucks in 1962. Opening its first integrated automotive assembly plant in 1973 in Korea, the company was knocked off its feet in 1981 when the new military dictator, Chun Doo-hwan, enforced industry consolidation. Kia was forced to give up passenger cars and focus entirely on light trucks. The company produced a few hundred cars in 1982 and 1983 before going dormant from 1984 into 1986 when it produced roughly a dozen. Kia rejoined the auto industry in partnership with Ford, producing ninety-five thousand cars in 1987 and growing to sales of 2.8 million in 2019.

Returning to VinFast, partnerships and investments have established the supply network; the key challenges to longer-term success are consumer Willingness to Commit and financing, or Ability to Profit. The past year (2023) has been a scramble to keep the company liquid. In April 2022, *Barron's* reported that the company planned a traditional IPO on a US stock exchange, despite the slow market. As of April 4, only 26 companies had listed shares, compared to 101 in the same period in 2021. As 2022 progressed into 2023, the outlook for IPOs only worsened.[9] On December 7, Joe White wrote an article for Reuters, saying,

> Oh my poor brain! . . . Let's make sense of it all. VinFast challenges the bears One thing is for sure: Vietnamese EV startup VinFast chose its name well. "Fast" is what this company has been all about since it showed its first cars in Paris just four years ago. VinFast faces a tough road to achieve its goal of a $60 billion valuation in a market that is nowhere near as welcoming of little-known, money-losing EV startups as it was in 2021, during the peak of the blank-check, SPAC deal boom.[10]

The fast-moving culture of VinFast operating in a tightly controlled communist country was discovering that a freer, highly financially regulated Western culture might be a bit different to operate in. Particularly given the high burn rate for its cash reserves.

By June 2023, Reuters was reporting that the company lost $598 million in Q1 2023, following $3.4 billion in losses the previous two years.[11] VinFast leaders sought to characterize this as due to the phase-out of ICE vehicles. Chairman Vuong's choice to dump traditional cars and go all in on EVs was bold, allowing the company to move down the learning curve of battery manufacturing quickly and hopefully develop a core competence that provides a strategic advantage, but it starved the company of revenues while facing a long trip across a desert. Earlier in the year, an announcement came that VinGroup and Chairman Vuoung were investing several billion dollars in VinFast. Vuong may be the wealthiest man in Vietnam for now, but he does not have pockets as deep as Elon Musk or Jeff Bezos, and the financing of ongoing initiatives appears to be becoming a critical threat to the company.

There is a saying that timing is everything, which appears to apply to Lucid Motors, which has been ahead of its Vietnamese competitor and managed to get to an IPO a critical half-year sooner. Founded in 2007 as a battery company called Atieva, Lucid changed directions in 2016, recruiting Peter Rawlinson—the former lead engineer of the Model S program at Tesla—to head a push to build its EV sedan. In 2018 Lucid found the deep pockets of Saudi Arabia's sovereign wealth fund in 2018, arranging a $1 billion deal in September of that year, just a few weeks before Crown Prince Mohammed bin Salman had Washington Post journalist Jamal Khashoggi brutally murdered. Including later investments by the sovereign wealth fund, Saudi Arabia obtained majority ownership of the company. In July 2021, the company went public in an IPA reaping $4.5 billion.[12]

Lucid made some substantially different choices with how it used this cash, choosing to offer a vehicle that lands firmly on the luxury end of the spectrum. While a 2023 Air Pure RWD model can be had for a little under $80,000 with a range of 519 miles, the Grand Touring climbs to over $120 thousand and the Saphire to over a quarter million dollars. The *Car and Driver* review was very positive, while clarifying that this is a luxury vehicle priced as such.[13] In August's earnings announcement, it was revealed that the company had lost over $750 million in Q2. However, revenue had increased to $150 million, up from $97 million the previous year. This increase was due to the production and sale of almost three thousand cars in the first two quarters, which does not look very different from VinFast.

Upon further digging, Lucid's competitive position entering Q4 of 2023 has several advantages over those held by VinFast. First, it took a different strategic approach and invested in facilities designed for lower-scale production of a luxury product. Consequently, Lucid's capital investment is lower, making break-even a lower threshold. Second, Lucid's financing was supplemented by another offering in May 2023, which brought in another $3 billion—an amount sufficient to extend its cash burn runway another year into 2025. Finally, the Kingdom of Saudi Arabia was committed to buying between fifty and one hundred thousand Lucid Airs over the next decade.[14] Even at the low end of $100,000 per vehicle, this represents a minimum of $5 billion and a high end of $10 billion in purchase commitments. A very nice position to be in.

Returning to VinFast, the difference between Lucid completing its IPO in July 2021 and moving toward an IPO for Vuong's company in April 2022 is gigantic. Combined with the additional challenges of a Vietnamese company penetrating the US market and strategic choices, the interest in supporting an IPO was much lower. In May 2023 came the announcement that VinFast would partner with Black Spade a special-purpose acquisition company (SPAC). In plain English, a SPAC is a kind of holding company in which investors buy shares, the money is held until a deal is formalized and the shareholders get to vote to approve the deal or have their investment returned. The announcement was that Black Spade would merge with VinFast in a manner that would place a value of $23 billion on the EV startup. Again in plain English, Black Spade was bringing a little over $150 million to the table, meaning that the SPAC would own a little over 0.5 percent or about one in two hundred shares. A good deal? A bad deal? Time would tell, as would a shareholders meeting held at 9 p.m. Hong Kong time (where Black Spade is registered), or 9 a.m. Eastern Standard Time. As a small shareholder, I was very curious to see what would unfold.

At the appointed time, Kester Ng, the co-CEO came on the call to be the master of ceremonies, reading a short legal-sounding statement. The proposal for the meeting was not whether to merge with VinFast; rather, the directors of Black Spade had made a motion to extend the deadline to find a deal for the SPAC by twelve months. As the Mr. Ng read the prepared statement at the Hong Kong jockey club, longtime home of horse racing and big deals for the former British colony, he said that shareholders had either voted their shares by proxy or there would be a brief voting period. Interesting—I had only bought my shares a few weeks earlier and was not in time to be a registered voter, so I was attending out of curiosity. The voting period took at most thirty seconds. After which it was announced the motion had passed with no mention of the vote counts. I thought, well, that was a very strange meeting.

Not having time to inquire further that day, I woke up the next morning to a string of WhatsApp messages from a correspondent for one of the major news organizations. He and I had been comparing notes leading up to this meeting—his texts were excited and sarcastic, as if a jewelry heist had occurred. The reporting from VinFast itself was much more buttoned down:

> In connection with the vote to approve the Articles Amendment, the holders of 14,150,715 Class A ordinary shares of the Company properly exercised their right to redeem their shares for cash at a redemption price of approximately $10.38867176 per share, for an aggregate redemption amount of approximately $147.01 million. Following the redemption in connection with the Articles Amendment, the amount of funds remaining in the trust account is approximately $28.56 million.[15]

Reading between the lines, six out of every seven shareholders had voted to run from this deal, roughly 83 percent preferred getting their cash back to investing in VinFast. If Butch Cassidy and the Sundance Kid had been able to move that fast after their last heist in Bolivia, they would have lived a much longer life.

I turn back to my interview with Robert Ligon, one of the first Western automobile experts to join DeLuca and the VinFast team. He signed on with the team, receiving some supportive advice from DeLuca. Robert reports that his initial contract offer did not have a clause regarding compensation if he was terminated and that DeLuca advised him to ask for that clause before signing, which he did and received one. VinFast and Chairman Vuong have a low tolerance for people they don't think are in the correct position. Madame Thuy and the highest-level executives push people very hard. Very few people have achieved Vuong's success without such aggressiveness. Elon Musk, Steve Jobs, and Jeff Bezos are all known, charitably, as being difficult bosses. Jim DeLuca was a critical component to building and making VinFast possible, but the buck did not stop with him, thus he worked on developing methods to cushion the inevitable major bumps in the road.

DeLuca related to me that on numerous occasions Madame Thuy would come into his office and express dissatisfaction with a high-level employee. The conversation would explore opportunities for improvement but usually ended with the command that employees needed to be fired—immediately. So Jim developed a technique:

> I began taking a single employee out to dinner and everybody caught on. They knew if Jim was asking you out to dinner, you're gonna get a severance package. The first time I did it was with an employee who had come out of General Motors with a tremendously successful record and was a great, great leader and very good at process, but he didn't have the ability to move fast and so the upper leadership quickly became frustrated with him. So I sat down with him and another employee and I told him, "Hey, it's really not working and we're probably going to have to part ways. And here's the offer." He said, "Well, okay, how long do I have to make a decision?" I stole the line from Don Hackworth at General Motors when he was talking to Gary Cowger about a move. I said, "Do you see that bottle of wine on the table?" He said, "Yeah." I said, "We can drink it fast, or we can drink it slow, but when it's done, I need your answer." He said, "Okay, I get it. I'm gone." And he left happily.[16]

This employee was also impressed by the ability of VinFast and its leadership to build a competitive automotive manufacturer from nothing. He discussed the strategic partnerships with suppliers and BMW, which he had been part of developing, noting the many strong accomplishments. At the same time, he offered his own thoughts on the Vietnamese and VinFast culture, which may be tinged by how his tenure in Hanoi ended. Notably Ligon said: "These people were honorable, but crazy. They were also arrogant because they were led by Vuong, who had made all of his money in Vietnam. Going global is a very different ball of wax."[17] When Ligon was ultimately fired from VinFast, DeLuca advised him to demand "double termination," which he received; in addition, he and his wife were treated to two weeks at one of the chairman's VinPearl resorts, including business class airfare from the United States.

Since the turn of the century, all VinGroup operations have been in Vietnam, which, as a communist country, operates differently than the United States or Europe. One quick example occurred just before Christmas 2022 when the provost and president of VinUniversity presided over the VinFutures prize ceremony. The grand prize of $3 million was awarded in a ceremony at VinUniversity in Hanoi to Sir Timothy John Berners-Lee, inventor of the World Wide Web; Dr. Vinton Gray Cerr and Dr. Robert Elliot Kahn, who led the design of TCP/IP; and Dr. David Neil Payne, who had conducted much of the early work that developed optical fiber communication. I talked to two faculty members at Vin University who independently told me that most of the people in the crowd were watching Prime Minister Pham Minh Chinh and Chairman Vuong very closely, wondering what their relationship looked like, who was more powerful, and in what ways.

As autumn 2023 approached, a batch of news stories appeared that suggested VinFast had overshot its ambitions. First, Nikkei published an article titled "Vietnam Vingroup's Ambition to Take on Tesla Hits Bumpy Road." VinFast had sold just 7,400 automobiles in 2022, which amounts to a utilization rate of just 3 percent in its Haiphong facility. Nikkei offered a rare public quote from the chairman, saying, "By 2024, sales will reach 60,000 to 70,000 units, and EBITA will break even. We will be able to generate profits in fiscal 2025."[18] Nikkei concluded the outlook was murky. A quick back-of-the-envelope calculation suggests the outlook was worse than murky. Taking the estimated production and multiplying it by an average sales price of $50,000 yields projected revenue of $3.5 billion for VinFast in 2024.That might come close to break-even but will not cover the current burn rate for cash—and it assumes that the cars can be sold, which is a stretch assumption given the reviews they have received so far in America.

In another article, it was reported that VinFast had reached out to and even pressured North Carolina governor Roy Cooper to lobby the US DOE to approve a $1.4 billion loan for VinFast. The company's vice president of government relations and strategic partnerships, Brook Taylor, contacted the governor's office several times, writing in an email April 13: "Unfortunately, the process seems to have slowed down as the DOE focuses attention on projects that will send federal funding to other states." And with mounting frustration:

> The state of North Carolina has stepped up and provided substantive support for the project. Meanwhile the (Biden) Administration has taken credit for the efforts of VinFast and North Carolina—while eliminating our vehicles from federal incentive eligibility, raising interest rates and making it harder to access private capital, and dragging out the process for a federal loan.[19]

Governor Cooper did make a call to Jennifer Granholm, secretary of the DOE, but he did not seem to be heavily vested in this lobbying effort. Welcome to America VinFast. The governor was perfectly happy to have his picture taken on March 29, 2022, when the plans for a plant in North Carolina were announced. Cooper was also front and center when VinFast broke ground on its "Crown Jewel" on July 28, 2023.[20]

At the same time, the news out of Europe was not any more encouraging. *Automotive News Europe* reported on August 3 that VinFast was dropping out of the Munich Auto Show to be held in early September. While no specific reason was provided, Chairwoman Thuy stated at the Automotive News Europe Congress in June that the company's planned launch in Germany, France, and the Netherlands had been slowed by more challenging safety regulations. The implication was that VinFast would stake its future on the ability to sell cars in the United States.

The news through the first eight months of 2023 strongly suggests that Chairman Vuong and VinFast may have stretched too far too fast. A common principle of warfare is to fight on only one front as throughout history armies have become bogged down and defeated by being stretched too thin. It is a reasonable argument that the only successful war prosecuted and won on two major fronts was the United States in Europe and Asia/Japan in World War II. The foundation for that success was all economic, as numerous historians have discussed how the American economy was jolted out of the Great Depression, and between Pearl Harbor and the end of 1943 American industrial might was able to outfit the US Army, Navy, and Air Force.

Simply put, VinFast needs a large capital infusion as well as a strong signal of credibility and viability. At the groundbreaking ceremony for its North Carolina plant on July 29 the company announced that the US

Securities and Exchange Commission had announced the validation of its F-4 filing, denoting a step forward in its proposed merger with Black Spade. VinFast's global CEO, Madame Thuy Le, said, "Today represents a remarkable milestone in advancing VinFast's presence in the U.S. as we move towards our proposed U.S. listing. Along with this exciting step, today we also celebrate the start of construction of our electric vehicle (EV) factory in North Carolina."[21] Madame Thuy can express all the optimism she wants, but unless Black Spade has had a major capital infusion since July when 80 percent of its stockholders walked away, the $30 million in this offering will only cover one day of operations.

In a persistent effort to provide a carefully curated story of positivity, one of the photographers invited to the North Carolina groundbreaking was Huỳnh Công Út, known professionally as Nick Ut. Ut won the 1973 Pulitzer Prize for his iconic photograph of children running from a South Vietnamese napalm attack that hit their village in 1972. Ut noted, "Two years ago, I hear that there's a Vietnamese car, I didn't believe it (until he traveled to the Haiphong factory). I went back to Vietnam after the war. The country had nothing left."[22]

On August 15, the company debuted on the NASDAQ under the symbol VFS. Showered with confetti, a group of fifteen VinFast executives rang the opening bell with Deputy CEO Michael Johnson, a thirty-year veteran of Ford, and Thuy Le, CEO, standing front and center. A triumphant moment, at days end the newly listed corporation was valued at $65 billion—more than either of the iconic American companies Ford and GM. To justify this exorbitant valuation, the company will need an infusion of several billion dollars. Like Mr. Ut, I am pulling for his home country and VinFast while being fully cognizant of the extremely difficult journey it is facing.

Most-Recent Update

Since the writing of this chapter, there has been a steady drumbeat of VinFast updates and news reports. In mid-August, the valuation of the company on the Nasdaq briefly eclipsed the value of General Motors and Ford combined, before rapidly falling to approximately 10 percent of that value. Meanwhile mainstream news outlets including ABC news,

Reuters, the *Wall Street Journal*, and the *New York Times* started finding stories to write about the company. Worldwide, the company sold just 19,652 EVs in the six-month period ending September 30. At the same time, the company under CEO Le Thi Thu Thuy's leadership has shown remarkable resilience and ability to make strategic shifts. This includes continuing to compete in the United States as the most difficult market in the world, where despite selling 2,009 vehicles in 2023, Thuy says, "We wanted to go make our name in a very difficult market. Our rationale was very simple. If we can make it there, I mean, people will believe in us. So it's an approval stamp to some extent. But it is very difficult."[23]

Additional moves by VinFast include seeking to enter new markets, such as Europe, Thailand, and India, as well as arranging a $1 billion financing and cash infusion from Yorkville Advisors at the end of October.[24] This cash infusion should provide the company with a longer runway to prove itself, with two strategic moves of interest. First, the company recruited twenty-seven potential automobile dealers across twelve states to enhance its original direct-to-customer sales model. Second, the company floated a design concept for a VF 3 minicar specifically designed for the Vietnamese market that could have potential as a city car for budget-minded Americans.[25] With these moves the company founded by Pham Nat Vuong was in the fight, perhaps looking a bit like Muhammed Ali in the third round of his "Thrilla in Manila" fight in 1975. In that round Ali used the "rope a dope" strategy to tire Joe Frazier out before emerging as the victor at the conclusion of the fourteenth round. Could Vinfast win its fight? Possibly?

Chapter 6

Reinventing General Motors

Amid the ignition switch recall discussed in chapter 4, President Barra and GM were fighting to survive, yet as 2014 progressed Barra began the pivot from the old GM to the new GM. In September 2014, she participated in a Clinton Global Initiative plenary session in conjunction with Climate Week in New York. In a discussion moderated by Chelsea Clinton, Barra elaborated on several themes she sought to lead the company with:

> *Regarding the Car Industry's Future*: "As you look at the future of the auto industry, I believe it'll change more in the next five to ten years than it has in the last thirty to forty to benefit society. Look at the connectivity, vehicle-to-vehicle communication, and vehicle to infrastructure, which makes the cars fundamentally safer and reduces congestion, and then also from an environmental perspective there is tremendous change happening in the auto industry. But it'll only work if it's collaborative. We have to do it together to get the maximum benefit."
>
> *Regarding Sustainability Driving Business Value*: "There's so much we can do with technologies, whether it's electrification, whether it's more fuel-efficient internal combustion engines, whether it's the connectivity that'll change the way that we can attack safety, congestion, and fuel economy. It all comes together. It starts with a customer expectation, but it's a fundamental part of business."

> *Concerning doing the Right Thing*: "When you look at the world being more connected, the world has become a lot smaller. It also has provided transparency of information so that everybody has higher expectations, as they should. Whether it's a small company or a large company, there is an obligation. With opportunity comes the responsibility to do it the right way."[1]

Here we see a leader settling in and starting to look to the future with a long-term vision. While her estimate that there will be more change in the next five to ten years was a bit off, the remainder of this chapter profiles her efforts to lead the vast GM team toward a greener, more equitable future with Jim DeLuca as a critical partner.

According to DeLuca, a key move was taking the senior leadership team to California in 2015. As Jim remembers it:

> We went to San Francisco because she said we need to look at how the fast movers are working, what they're doing. If we keep doing what we've always done, we're not going to be here. And so we went out to California.
>
> We spent time with Apple, we spent time with Google, we spent time with Cisco. This was a period where we had already moved. I think we'd already purchased Cruise.
>
> And this is when Mary started to develop the concept that we need to move in a different direction. And that was back in 2015.[2]

While DeLuca's memory is imperfect (the Cruise investment was made in March 2016), the key point regards the investment in the future and the fortitude to set stretch goals. The team that went included nine officers of the corporation and four senior VPs. Combining travel costs, time away from daily operations, and the coordination involved in setting time up with tech companies, this was quickly a million-dollar investment.

According to DeLuca:

> So Mary had the intestinal fortitude to do two things. Plot a new path and get out of businesses that weren't returning a profit. So just leave

> Russia, leave Thailand, leave Australia, and eventually leave Europe. Let's face it, Fritz Henderson [CEO March–December 2009] wanted to leave Europe and the Board said no, instead they got rid of Fritz. A few years later, Mary said, we got to get out of Europe. After we spent billions. We could've saved if we had left when Fritz wanted to leave.
>
> So I give Mary credit for both normalizing the business and having the vision to say, *we're going to move in this new direction*.[3]

In September 2017, Barra hosted a press conference in Shanghai to outline a roadmap for safer, better, and more sustainable transportation solutions. In discussing electrification, she noted that "Our engineers have continually built on our experience" and cited the Chevrolet Bolt EV that was introduced in the United States in 2016. With a range of almost four hundred kilometers and nearly 45 million kilometers on the road at the time, the future looked bright. At the time, GM planned to roll out ten new EVs in China between 2016 and 2020. Two things can be true about one statement. First, this was a very forward-looking, ambitious statement. Second, it was not highly accurate. The Bolt has been widely discussed as GM's first substantive foray into EVs, but this ignores an earlier effort with the Spark, which Kevin Johnson helped lead in Korea in 2009. I will come back to that later. Another thing to note from the Shanghai press conference is the significant investment and grand plans for Chinese production and sales. In short, these have not panned out well for GM or other legacy Western manufacturers. A strong argument can be made that these efforts helped Chinese companies move down the auto production learning curve and assimilate knowledge that has significantly contributed to the present-day lead China owns. This book is focused primarily on the US market; thus I will set that aside while noting that looking at auto production as a local market is very shortsighted. Nonetheless, the 2017 press conference shows President Barra advancing a bold vision.

During the two-plus years the Prince and Queen worked hand in hand before DeLuca retired from GM (announced June 10, 2016), the pair advanced GM's dual aims of profitability and greater sustainability, with Barra being far more visible. In December 2015, DeLuca took the lead in announcing a significant investment in wind energy in Texas. GM committed to purchasing 30 MW of clean electricity from the Hidalgo wind farm in Edinburg, Texas. In noting the benefits of this purchase agreement for the Arlington GM plant that produced 1,200 vehicles per day, DeLuca said: "Our sustainable manufacturing mindset benefits the communities in which we operate across the globe." Projections were that the Arlington assembly plant would cut around $2.8 million in annual energy costs and reduce one million metric tons of CO_2 emissions during the period of the fourteen-year deal.

Flashing forward to the July 2023, Texas has been baking under a scorching heat wave, and numerous news sources and social media posts have highlighted the critical contribution of renewable power in keeping the electric grid in Texas up and running.

On July 6, in Dallas, midway through a weeklong stretch where the average maximum temperature was nearly 97°F, and the maximum daily temperature was 100°F, solar energy produced 10,700 MW, while wind generated 7,400 MW, or about 26 percent of the grid's power. GM's investment in 2015 was an early indicator of willingness to commit long before the Biden administration passed the IRA. In short, early movement to renewables, partly spurred by GM, stabilized a frazzled and frayed power grid.

In June 2016, GM announced a record calendar year income for 2015 of almost $10 billion. Mary Barra noted:

> It was a strong year on many fronts, capped with record sales and earnings, and a substantial return of capital to our shareholders. We continue to strengthen our core business, which is laying the foundation for the company to lead in the transformation of personal mobility. We believe the opportunities this will create in connectivity, autonomous, car-sharing and electrification will set the stage for driving value for our owners for years to come.[4]

At the time, Tesla's market capitalization was almost $50 billion, while GM's was $42.7 billion. Chapter 3 examines the rise and evolution of Tesla, while the remainder of this chapter delves into GM's investments and strategic moves regarding electrification from this point forward.

Dialing Up Willingness to Commit on Electrification

One of the challenges in representing history, particularly for a storied organization such as GM, involves sorting through an ocean of diverse data, memories, and historical reports. Earlier, I cited Mary Barra's effusive discussion of the Bolt, widely considered to be GM's first major EV effort. Yet, in discussions with Kevin Johnson regarding his role in the early stages of VinFast, he mentioned the Chevrolet Spark, which he worked on during his time with GM in Korea. Originally developed as a city car (i.e., small, fuel-efficient and not particularly sporty) in 1998, GM announced in 2011 that it would produce an EV version of Spark as a compliance car to meet California's tightening emissions targets. The Spark was the first all-electric car marketed by GM since the EV1, which was discontinued in 1999. The Spark EV had a 97 kW electric motor and a range of eighty-two miles, with a MSRP of $25,995. It did not sell well, with total sales of a little over seven thousand units.[5] In this case GM's willingness to commit was solely based on government regulation—no one thought the car would make money.

Beginning her third year in the CEO seat, Barra and GM made a steady stream of announcements and investments throughout 2016 signaling a more tech-focused strategy. Following a $500 million investment in Lyft in January, the company announced in March that it would buy a forty-person startup, Cruise, with an investment estimated at $1 billion. A company working on developing hardware and software to be installed in original equipment—that is, traditional cars—to enable self-driving, the purchase of Cruise was praised by Venture capitalist Nabeel Hyatt, who said, "They moved faster than most Silicon Valley companies would move."[6]

In October, the company began production of the Chevrolet Bolt at its Lake Orion plant in Michigan with the battery pack and drivetrain components supplied by LG Corporation. A fascinating test drive and review of the Bolt was offered at the time by Brian Fung in the *Washington Post*. One observation regarding the MSRP of $37,495 was that this made it one of the most affordable electric cars around in a nod to the mainstream consumer. Fung wrote,

> The Bolt's mechanical simplicity means its floor can be built completely flat, creating extra space that leaves the car feeling roomier on the inside. This feeling is only enhanced by the larger, taller windows that give the driver a wider field of view. Performance-wise, the Bolt is about as peppy and responsive as any car you've probably driven, if not more. That's thanks to the nature of electric driving: Unlike conventional engines that need to rev up, electric motors can apply maximum torque instantly. The Bolt can accelerate from 0 to 60 in 6.9 seconds, according to GM. That's comparable to the 2016 Honda Civic 1.5-liter turbo, which accomplishes the task in 6.8 seconds.
>
> It's still unclear whether many consumers will warm to the Chevy Bolt. But it will have some lead time on its competitor, the Tesla Model 3, allowing Chevrolet to make an early first impression. And while the Bolt makes some clear departures from other, more traditional cars, Chevy seems to be hoping that it will seem familiar enough for people to consider adopting it.[7]

With these good reviews and a lead on Tesla, the Bolt, the first mass-market all-electric car with a range over two hundred miles should have sold well. Only it didn't, totaling sales of about one hundred eighty-six thousand units in six years before the announcement in early 2023 that GM was discontinuing the model. For comparison's sake, in 2021, GM sold over half a million gasoline powered Chevrolet Silverado trucks at a much higher average sticker price.

Which brings me back to the Hummer, the ICE version of which was profiled in chapter 2 with Arnold Schwarzenegger as the celebrity first driver in 1991. In January 2020, the company announced it was bringing back the Hummer as an EV. With assembly in Detroit and batteries produced by LG Chem in South Korea, this product was intended to be electric and produce lower emissions like the Bolt, but with a much higher profit level. The base version has 830 HP and is rated to tow up to 7,500 pounds. With a 3,000-pound battery, the Hummer is so large that it is rated as a class 3 medium-duty truck in the United States. As a comparison, the eighteen-wheel semi-trailers on US highways are class 8 trucks. With a base MSRP of over $100,000, this is a car designed to enhance GM's profitability, not to be a mass-market product. The car has been marketed with GM's usual flair and pizazz, including a commercial that debuted a feature named "crab walk" that debuted in October 2020 and became a very popular feature; there are videos of crab walk in action on YouTube that have over 11 million views.

Automotive News reported in March 2023 that GM had over ninety thousand reservations for the electric hummer. Duncan Aldred, VP of global Buick and GMC said "We had to pause those reservations because they were coming in so quickly we simply couldn't keep up with them on a production level."[8] This is a vast understatement, with *InsideEVs* reporting on July 5 that GM only delivered forty-seven vehicles to customers in the second quarter.[9] This revolution is not easy; in this case customers appear to be willing to commit, but GM has not yet transformed its supply network sufficiently to meet what is hoped to be strong consumer demand.

As a final signal of GM's willingness to commit, in January 2021 a new logo was introduced to replace the iconic all-caps white lettering on a blue square that had been employed for half a century. The new logo is intended to evoke "the clean skies of a zero-emissions future" as well as the "energy" of its Ultium battery platform technology. In the tradition of all great logos, there is also a subliminal message with the negative space in the "m" forming the shape of an electrical plug.[10]

Transforming the Supply Network—Partnering with Honda and Others

As far back as 2018, GM and Honda were finding ways to share an industry's risks and potential payoffs in transition. In October, the *New York Times* reported that Honda invested $750 million into Cruise to obtain a 5.7 percent stake. This announcement signaled a broad-based approach to developing self-driving cars, with both GM and Honda claiming the other brought key capabilities to the table. In particular, Honda has an expertise in compact vehicles that should help Cruise. In addition, GM brought in SoftBank, a Japanese tech investment fund in early 2018, with a $2.25 billion investment.

Another partnership between the Motor City Giant and Soichiro Honda's company was announced in September 2020. This followed an April announcement that the two companies would jointly develop two all-new EVs for Honda. The September announcement focused on combining vehicle development, with specific vehicles to be sold or badged under both labels. In other words, the same car would be sold as a Honda and GM, albeit with some styling and interior differences. In addition, the two giants planned to collaborate on purchasing supplies to leverage combined scale and cost savings. According to Seiji Kuraishi, executive VP of Honda:

> Through this new alliance with GM, we can achieve substantial cost efficiencies in North America that will enable us to invest in future mobility technology, while maintaining our own distinct and competitive product offerings. Combining the strengths of each company, and by carefully determining what we will do on our own and what we will do in collaboration, we will strive to build a win-win relationship to create new value for our customers.[11]

In October 2021 GM announced it was building an electrical vehicle battery lab on the grounds of its Technical Center. Tim Grewe, GM's director of battery cell engineering and strategy put it simply, saying, "We need to make better batteries that cost a *lot less*."[12] The announcement noted that GM was spending hundreds of millions of dollars on the lab, but in supply networks and car manufacturing, the devil is absolutely in the details. Consider the Chevy Bolt and its supply network evolution as bookended in a pair of *Wall Street Journal* articles. In October 2015, a joint announcement noted that LG Chem would provide eleven systems for the Bolt, including the electric-drive motor, inverters, power control module, heating/air-conditioning system, and the entertainment system. LG Chem was pushing deeper into supplying the auto industry as Apple was readying its electric car prototype. Ken Chang, an LG vice president, said, "Our vision is to be a component supplier for electric vehicle manufacturers."[13]

Turning the clock forward six years, another article in the *Wall Street Journal* was startingly negative. In this announcement, GM announced it would recover almost $2 billion from LG for manufacturing defects in the batteries and equipment LG had supplied for the 142,000 Bolts manufactured in the half-decade since the car debuted in 2016. GM had claimed the fault likely stemmed from a manufacturing problem at LG factories that resulted in two flaws in specific battery cells that could lead to a fire. *At the time of the announcement (2021), GM had been telling owners of Bolts to move their cars outside immediately after charging, not to park the car inside overnight, and to leave abundant distance from other vehicles in parking garages.*[14] A setback for GM and LG? Absolutely! At the same time, worth comparing. Recall from chapter 1 that Henry Ford incorporated his entire company for the equivalent of $900,000 in 1903 and that the Model T was his *twentieth* car model. Failure, persistence, and improvement may lead to ultimate success. In the case of the GM/LG partnership, the settlement to GM amounted to roughly $13,000 per car sold. Thus about sixty-seven recalled Bolts represented the same dollar amount as Henry had invested in founding Ford Motor Co.

The scale of investment and change today is orders of magnitude higher than with the emergence of ICE vehicles in the early nineteenth century. Mary Barra-led GM is undeterred by setbacks such as the LG Bolt battery problems. The company is forging ahead with vertically integrated and deep supplier tie-in investments, including plans to open a fourth battery plant. Sham Kunjur, executive director of GM's Raw Materials Center of Excellence, says the company is investing heavily in owning its supply network or heavy investment with key suppliers because "that is what we needed to help us grow this market." Sam Abuelsamid of Guidehouse Insights interpreted the planned investments, saying GM "has absolutely been more aggressive than any other automaker aside from Tesla" in terms of vertical integration. GM's Kunjur noted that "What was available in North America, it wasn't much, and we didn't see a lot of movement in this space by natural forces. We felt we needed to control our own destiny."[15]

One fundamental factor underpinning GM's strategy is China's huge lead in the EV supply chain. Auto manufacturers worldwide are desperately seeking sources of raw materials and batteries closer to home. A second fundamental factor was Tesla's strong partnership with Panasonic. In short, absent an alternative, GM felt it had to work with LG and forge a path through inevitable stumbles and quality challenges.

The first battery plant owned and operated in partnership with LG Chem opened in August 2022 near Lordstown in northeast Ohio. The GM-LG Chem joint venture, Ultium, has two more battery factories planned in Tennessee and Michigan. At the same time, a fourth plant was announced in mid-June 2023 by Indiana governor Eric Holcomb. Together, these four battery plants represent an investment of several billion dollars and a chance for GM to establish itself as a clear second place to Tesla relative to Ford, which has yet to open a single battery plant at this time.

Electric vehicles rely heavily on batteries and powertrain components, which are considered Tier 1 elements. However, General Motors (GM) has taken a step further by securing supply from Tier 2 suppliers. In 2021, GM made a deal with Controlled Thermal Resources to extract lithium from geothermal brine in southern California. This method, known as "direct lithium extraction," is different from the South American extraction process that involves ponds and generates tailings. GM's decision to secure a lithium supply is due to concerns about access to this crucial mineral. The mining becomes a "closed loop system. . . . It's set up to be a much lower cost supply chain" according to Tim Grewe.[16]

Five months later, GM announced another partnership with Posco Future M to obtain cathode active material (CAM) a key battery material representing 40 percent of the cost of the cell. GM later expanded its investment in its Quebec supplier to $1 billion. Notably GM does not fully own or control any of these ventures, nor is it a pure purchase agreement, since GM is providing startup capital. This places these ventures somewhere in between the classic choices of Make or Buy on which purchasing strategy typically anchors. This is because GM is both investing in the companies and committing to purchase contracts.

In another dramatic deal, GM announced in January 2023 a $650 million investment in the Thacker Pass lithium mine in Nevada. The deal with Lithium Americas Corporation allows GM preferential access to sufficient lithium to build up to 1 million EVs annually. On the day of the announcement, shares of Lithium Americas rose 14 percent, and GM's stock rose 8.4 percent. In a speech to the US Senate, Joe Manchin crowed that the investment was a "tangible result" of the Inflation Reduction Act passed in 2022.[17] According to the GM's Kunjur, "If you'd asked us three years ago or four years ago if we would be directly engaged with mining companies, we would have clearly said no, but sometimes necessity is the mother of invention."[18]

In addition to the sourcing commitment, GM agreed to buy $650 million in shares in Lithium Americas in two equal payments, with the first payment only coming if Lithium Americas prevails in a long-running court case. That court case is in US federal court and hinges on whether former president Trump erred when he approved the Thacker Pass lithium mining project in northern Nevada. The project is an innovative application as it seeks to extract lithium from a large clay deposit, which has never been done at a commercial scale before. The stakes in this court case are monumental. On September 9, 2023, the market capitalization of Lithium America's stood at a shade under $10 million. A company that GM had signed a contract to invest over a half billion dollars in and had agreed to buy all of the lithium produced in the mine when it begins production (projected to be 2026) would be expected to be valued at over half a billion dollars. So why the low valuation?

The low valuation is because of the legal uncertainty for this company. On July 17, a three-judge panel in the San Francisco appellate court rejected a half-dozen legal arguments from a group of conservationists and tribal leaders opposing the opening of the mine on US federal lands.[19] Success! Lithium America's stock surged, correct? No, this case is likely headed to the US Supreme Court. Prior to the appellate court ruling, Roger Flynn, the director of the Colorado-based Western Mining Action Project, argued, "The district court made a very serious error here in not vacating this decision in the face of massive environmental damage and the serious errors that the Bureau of Land Management now admits happened here."[20] Opponents of the mine argue that it threatens the greater sage-grouse, an endangered ground-nesting bird that would lose thousands of acres of its prime winter habitat. The project would also threaten thousands of acres of pronghorn antelope range. Thus, it is a fair bet that the environmentalists opposing the mine have sufficient resources that their legal team will seek to have this case considered by the Supremes.

Lest the reader assume that all of GM's supply network investments are in North America, it is essential to note that the company also has invested heavily in China. In fact, China is also a critically important market as the company sold nearly three million vehicles or nearly half its total sales in 2021. In May 2022, the first all-electric LYRIQ vehicle rolled off the production line of the Shanghai Automotive Industry Corp.–GM joint venture factory in Shanghai. The joint venture was originally launched in 1997 with both SAIC and GM having an equal stake. The LYRIQ is Cadillac's first fully electric vehicle and the first car made using its proprietary vehicle platform Ultium.[21]

While I have focused this book on the North American market to enable feasibility and digestibility, players in the auto industry must monitor the worldwide market carefully, starting and beginning with China. In a highly salient opinion piece published in the *New York Times* on July 17, 2023, Robinson Meyer made a strong case for why America (and other countries) "can't build a green economy without China." Beginning with one of the OG automotive giants Henry Ford, there is a long history of innovation flowing across continents. In the early twentieth century engineers from Germany, France, Japan, and the Soviet Union traveled to the River Rouge plant to learn from Ford's methods. Similarly, Germany possessed the world's greatest chemical industry and knowledge in the early twentieth century, yet it wasn't until after the horrors of WWI and Germany's defeat that American companies, including DuPont and Dow, brought German scientists in to advance the American chemistry industry.[22]

The essence of innovation in industry after industry is what social scientists call "tacit knowledge" or know-how. Another illustration of the importance of know-how occurred in the 1980s when the Reagan administration pushed Japanese automakers to jointly build factories with American counterparts. The NUMMI partnership between GM and Toyota discussed in the Tesla chapter offers a case in point. Here GM associates began to understand and incorporate the key elements or DNA of the Toyota Production System—or lean production.[23] The automotive supply network is truly global and incredibly interconnected; innovation and tacit knowledge are everywhere; thus it is imperative for companies to continually scan and incorporate knowledge from a worldwide view.

Turning the Corner on Ability to Profit?

The Gears of Change are turning for GM, sometimes forward, sometimes backward. This brings me to the third gear, *Ability to Profit*. In addition to the well-known consumer tax credits embedded in the IRA (up to $7,500 for purchasing a new EV if all conditions are met) the act of Congress and the Biden administration also included substantial production subsidies that allow carmakers and battery suppliers to earn up to $45 per kilowatt hour. Tesla and Panasonic are jointly eligible for up to $1.8 billion in 2023, while the GM and LG Chem partnership is estimated at $480 million. Henry's company is not suitable for any subsidies until at least 2025. In aggregate, GM is building or operating plants estimated to have a capacity of roughly 125 gigawatt hours per year. This is about *three and a half times* Tesla's capacity in North America. In sum, GM is in the race. According to Abuelsamid of Guidehouse Insights, "They can probably catch up and surpass Tesla's North American volumes. . . . But it all comes down to doing it."

Just do it. Simple motto for Nike, yet substantially more complicated in the car industry. The Gears of Change all have to turn at the same time with little friction. As this chapter has examined Barra's leadership of GM into the EV era, some signs point to the gears starting to turn—for several companies including GM and Ford in addition to Tesla. On July 17, 2023, news came of a huge price cut in the price of the Ford F-150 Lightning. At the same time Ford announced that it had temporarily shuttered its River Rouge production facility so it could complete final production tweaks to enable an output rate of 150,000 trucks per year. The Rouge complex began construction in 1917 and was completed in 1928, becoming the largest integrated factory in the world. Then as now, scaling production from a few cars to many thousands per year is nothing like flipping on a light switch. Between May 26, 1927, when Henry and son Edsel drove the fifteen-millionth and last Model T off the line and December 2 when the first Model A was delivered, output at the Rouge was zero vehicles. For over half a year, Ford Motor had essentially zero revenue. A century later the situation looks oddly familiar for Ford, GM, Honda, and many others. Scaling up from a few thousand units a quarter or year to production volumes exceeding six digits is a complex process that can take a year or more.

In the second quarter of 2023, Ford sold almost forty-five hundred Lightnings, up almost 120 percent over the two thousand sold in all of 2022. The limiting factors? Both supply and demand. According to Tim Bartz, internet sales manager at Long McArthur Ford in Salina Kansas, fires in the Rouge that delayed production did not deter buyers so much as price. He elaborated that of the 135 reservations he had received from customers about 40 had canceled, saying, "Ford advertised a $40,000 electric vehicle and that attracted a lot of people. Now we've seen price increases and those people are like 'I'm out.'"[24] This clearly shows some problems with flagging interest in the Lightning, yet at the same time, some customers were paying six figures or more for an electric truck.

Returning to Ford's announcement on July 17, in addition to stating that it would soon triple production volumes, the company also announced price cuts of $10,000 putting some models at an MSRP of $50,000. In addition to the bottom-line price cuts, Ford also tendered an offer of a $1,000 bonus if customers build their own XLT, Lariat, or Platinum (the highest priced) models on the company website or dealer network before July 31. According to Ford Model e's chief customer officer Marin Gjaja,

> Shortly after launching the F-150 Lightning, rapidly rising material costs, supply constraints and other factors drove up the cost of the EV truck for Ford and our customers. We've continued to work in the

background to improve accessibility and affordability to help to lower prices for our customers and shorten the wait times for their new F-150 Lightning.[25]

Importantly, the price cuts also brought the entry level Lightning Pro under $50,000 thus qualifying it for up to $7,500 in federal tax credits under the IRA. This move is likely to generate a substantial bump in customer demand and increased orders. The stock market reacted by dumping both Ford and GM shares between 2 and 4 percent the same day while Tesla rose. While I'm not a financial expert, my thought is that the price-cut announcement is a strong signal that the legacy manufacturers are picking up speed in the transition from ICE to EV and beginning to turn a corner. That corner may only be first base, but it is still a corner.

Most-Recent Update

Since the writing of this chapter, Barra and GM have faced renewed challenges. In early October, the company took the Q3 US sales crown with the news breaking that the company sold a little over 650,000 vehicles, an increase of almost 20 percent from a year earlier. Second place Toyota came in lower at a little over 570,000. In EVs, GM sold slightly over twenty thousand EVs, mostly its oldest model the Bolt.[26] Overall GM had sold almost fifty thousand Bolts through the end of September, making it the company's bestselling EV ever by a wide margin. The company had announced in April that it planned to retire the model, as Barra told investors during an earnings call that "We have progressed so far that it's now time to plan the end of the Chevrolet Bolt EV and EUV production."[27] Reversing course three months later, Barra explained why the company was going to redesign the Bolt, adding its Ultium battery technology to enhance its features. In her explanation, she noted that this move would save billions in capital and engineering expense while simultaneously getting a new and improved product to the market two years faster.[28]

The moves GM made over the summer were very likely strategically tied to the immense challenges of transitioning from ICE to EV. Two particular challenges stand out. First, GM reached a tentative settlement with the United Auto Workers union on October 3, making it the last of the so-called Big Three (after Ford and Stellantis) to reach a settlement following a very expensive six-week strike that the UAW began in mid-September. The strike was nominally intended to claw back wage concessions that auto workers had made over the prior two decades. Another motive underlying the strike was a desire for workers to expand the union base—into the largely joint-venture battery production plants and also into nonunionized auto plants including Tesla, Toyota, Honda, and BMW. While Mary Barra said that she and other GM leaders were looking forward to having everyone back at work, UAW president Shawn Fain said: "We wholeheartedly believe our strike squeezed every last dime out of General Motors. They underestimated us. They underestimated you."[29] Barra and Jim Farley of Ford, likely breathed a sigh of relief while worrying about the added cost of enhanced wages, which Ford estimated at between $850 and $900 per vehicle.[30]

This brings us to the second substantial challenge facing GM, the reluctance of consumers to embrace EVs. It became increasingly obvious over the third quarter of 2023 that while EV sales were increasing at a substantially higher rate than ICE sales, the supply of vehicles was also backing up with ninety-seven days' supply of EVs on dealer lots versus a fifty-seven-day supply of internal combustion engine cars.[31] The reality that GM faced as I write this in early November is that early adopters for EVs have already adopted, talking the broader mass market into being willing to commit is a challenge, so while transforming its supply network the company led by Mary Barra also needed to protect its ability to profit. Keeping the gears of changing moving is difficult to say the least.

Chapter 7
The Nervous Giant

The Founder—1906

Soichiro Honda grew up helping his blacksmith father, Gihei, with his bicycle repair business near Hamamatsu, a city on the coast of Japan roughly 150 miles southwest of Tokyo. As a toddler, Soichiro fell in love the first time he saw an automobile in his village, repeating often later in life that he loved the smell of oil it gave off "like perfume." Demonstrating his creativity at a young age, Honda developed an ingenious method of hiding his poor performance in school from his parents. His school sent grade reports home with children, requiring a parent to stamp them with the family seal. Rummaging in his fathers' repair shop, Honda jerry-rigged a stamp to forge his family's seal from a used rubber bicycle cover. Unfortunately for young Soichiro, the stamp was supposed to be a mirror image. This worked fine for his family name, which is symmetrical when written vertically, but his friends' names were not so the forgeries he made for them were quickly discovered; thus, his forging scheme landed him in hot water.

Fortunately for Soichiro, his talents lay outside formal school, with him leaving home in 1921 to seek work. He found work as a car mechanic before returning home to open his own auto repair business in 1928. Passion for all things mechanical and fast led him to race in the first automobile race at Tamagawa Speedway in 1936. Driving a car that he had custom-built, Honda was unable to avoid a calamitous collision when another driver was exiting the pits as he passed—Soichiro and his brother, mechanic Benjiro, were thrown from the car, and his brother's spine was fractured, but Soichiro entered one additional race that October. He later recalled, "When my wife cried and begged me to stop, I had to give up."

Following his brief racing career, in 1937, Honda founded Tokai Seiki to produce piston rings for Toyota. The company developed into a thriving enterprise with a plant in Yamashita, which was destroyed in a B-29 bomber attack in 1944, while the Iwata plant collapsed in an earthquake in 1945.

Following the war, he sold the salvageable remains to Toyota for 400,000 yen (about $450,000 today) and founded the Honda Technical Research Institute in 1946. The company became the largest manufacturer of motorcycles in the world by 1959, all of which was based on Soichiro's passion for engines, as discussed relative to the Honda Effect in chapter 2. The company he built has always invested heavily in research and development while building expertise in internal combustion engines. The company first entered a car in the Formula One in 1964, just a year after first producing road cars. Winning its first race at the 1965 Mexican Grand Prix, by 1994 Honda was supplying engines to the CART IndyCar World Series, In 2004 Soichiro's brainchild won the Indianapolis 500 for the first time, providing full redemption for the failed racer from 1936. The founder passed away in 1991, but not before seeing his company expand to America and become seller of the bestselling car in America—the Accord—1989 through 1991.

HONDA COMES TO AMERICA

In early 1983, Joe Wall was lucky to sell three pounds of octopus a week in the Kroger in Upper Arlington, Ohio, where he managed the seafood department. Then, a customer mentioned that "with the addition of the Honda plant in Marysville, she told me there are a lot of Japanese in the area" and that there was latent demand. So she and Mr. Wall partnered to increase the seafood offerings and to label them in Japanese. The news about Kroger selling octopus spread quickly, resulting in weekly sales of twenty pounds. Additionally, the sales of squid, mackerel, shrimp, and whole black bass all increased by at least double. This story highlights two essential aspects of a successful business strategy, which I introduced in chapter 2 as the Honda Effect. First, great companies often have a core competence. For Kroger that is the ability to learn from its customers, as noted in the newspaper article from 1983: "The whole thrust of our business plan has been developed from consumer research."[1] Honda's core competence has been internal combustion engines. At the same time, there is also a strategic requirement to read the environment and adapt over time. While Honda has done that very well, it also has benefitted from strong support from its supply network and government. Thus, I step back in time to examine Honda's entry into the US market beginning in the mid-1960s and culminating with the company taking the coveted title of bestselling car with the Accord in 1989. This growth trajectory was not inevitable, just as the continued dominance of Tesla is not inevitable. At the same time, an examination may provide insights for other automotive manufacturers, including VinFast.

James Allen Rhodes was one of only seven US governors to serve sixteen years in office. First elected governor of Ohio in 1963, in his second term Rhodes sent National Guard troops to Kent State University in May 1970 to help control protestors demanding an end to the war in Vietnam. Crosby, Stills, Nash, and Young immortalized the events of May 4 singing about "tin soldiers," President Nixon, and four students dead in Ohio.

A powerful governor, Rhodes led the Ohio delegation that nominated Richard Nixon for president in 1968. Alas, the Ohio constitution limits the governor to two four-year terms in office; thus in 1971 Rhodes unwillingly retired. Far from quitting, he sued, and the Ohio Supreme Court ruled that the limitation was for consecutive terms, and he returned to office after narrowly defeating incumbent John Gilligan in the 1974 election.

Rhodes led the creation of a Transportation Research Center (TRC) when he proposed allocating $7.5 million in funds in 1965. That center was built and opened during Rhodes's exile from governorship. Opened in 1972, the TRC today is North America's largest multi-user proving ground and occupies 4,500 acres in East Liberty, Ohio, about forty-five miles northwest of Columbus. In operation for over fifty years, the TRC has always been self-funded and provides services to every auto manufacturing company with operations in the United States. The TRC would be instrumental when the once and again Governor Rhodes crowed on September 26, 1979, "Final discussions are underway with Honda officials, but no official announcement can be made until Honda reaches a decision." The newspaper article regarding Honda's future motorcycle plant mentioned that the company had selected a 247.5-acre site at the southwest boundary of the TRC and went on to write, "Last month the state provided $2.5 million for a site development for a Montgomery Ward center near Cincinnati and the state would use the same method for Honda."[2] At the time, investing in Montgomery Ward might have seemed like a good bet as its 1978 sales were $4.5 billion, yet by 1985, the company closed its catalog business after 113 years in operation. In 1977, Honda was Japan's largest motorcycle and fifth-largest automobile manufacturer, yet it wasn't to stay that way.

First, Honda had to acquire the land. The *Columbus Dispatch* broke the story the next day that Honda was negotiating to buy land next to the TRC that Governor Rhodes first promoted in 1965. The land Honda was interested in was owned by Ralph J. Stolle, an industrialist from Cincinnati and an acquaintance of the governor, who was in his third term

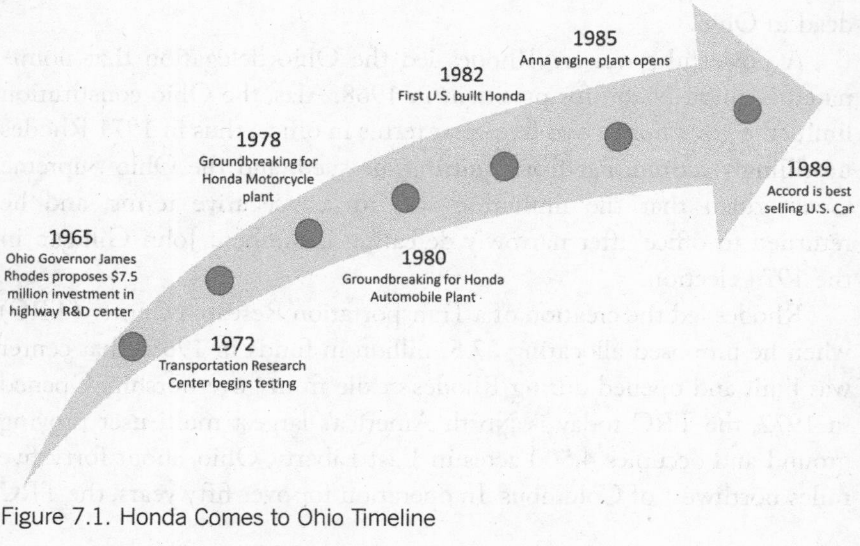
1965
Ohio Governor James Rhodes proposes $7.5 million investment in highway R&D center
1972
Transportation Research Center begins testing
1978
Groundbreaking for Honda Motorcycle plant
1980
Groundbreaking for Honda Automobile Plant
1982
First U.S. built Honda
1985
Anna engine plant opens
1989
Accord is best selling U.S. Car

Figure 7.1. Honda Comes to Ohio Timeline

in 1977. Lest anyone think there was anything shady going on, Rhodes noted that Stolle had bought the property in 1969 when the state of Ohio began acquiring 8,100 acres for the TRC: "They'll get that at no profit to anyone, and anyone else who wants to locate a plant there can get it [the property] at the 1968 price. They're going to sell it at what they paid for it. Any place you buy land in Ohio, people happen to be friends of mine."[3] Following the governor's spin, he was reelected for a final term in 1979, a year after Honda began constructing its motorcycle plant. This was followed by the groundbreaking for an automobile assembly plant in 1980, with the first American-produced Honda to roll off the line in 1982, and the opening of a separate plant focused on ICE production in Anna, Ohio, forty-four miles northwest of the motorcycle and automobile assembly plants.

As of 2023, Honda operates twelve manufacturing plants in the United States that produce 5 million products a year. The company was the first Japanese automaker in America to build engines and transmissions (1989) and to export cars to overseas markets (1987). In total, Honda has shipped over 1.4 million US-produced cars and employs twenty-three thousand associates. The governor who helped start this ascent ran for a fifth term in 1986 at seventy-seven but lost in a landslide to incumbent Governor Dick Celeste. While Rhodes did not win that election, his efforts to attract Honda were a huge success. He passed away in 2001 at the age of ninety-one, the same year Honda regained the top-selling title from Toyota. Today, the automotive giant with a core competence in ICE is nervous as it faces the need to transform its supply network completely.

Electrifying Honda 2023

In chapter 1, I profiled how Honda began with motorcycle production, developing a core competence in internal combustion engines and leveraging it to become the giant corporation of today. Thus, the announcement on October 11, 2022 (forty-five years to the day after Honda announced its first production facility in Ohio), represented a major revelation. The headline in the announcement was that Honda would spend $700 million to retool several of its existing auto and powertrain plants and an additional $3.5 billion in a joint venture with LG Energy Solution to build a battery plant in Fayette County, about forty miles southwest of Columbus. Bob Nelson, executive vice president of American Honda Motor Co. described the plans:

> Honda is proud of our history in Ohio, where our U.S. manufacturing operations began more than four decades ago. Now, as we expand Honda's partnership with Ohio, *we are investing in a workforce that will create the power source for our future Honda and Acura electric vehicles.* We want to thank the leaders of the state of Ohio, as well as in Fayette County, Jefferson Township, Jeffersonville, and Washington Court House, for welcoming this new joint venture between Honda and LG Energy Solution and giving us another Ohio community to call home.[4]

Much like Governor Rhodes a half-century earlier, Governor Mike DeWine wasted no time in ensuring that his name and political future were tied to the announcement.

> It has been more than four decades since Honda first saw great promise in Ohio, and although the way we manufacture vehicles is evolving, one thing that will stay the same is the quality of our workforce and their ability to get the job done. Honda and LG Energy Solution now join a long list of companies that have looked all over the country for the best place to do business and have chosen Ohio because we have the ideal economic climate and an innovative and talented workforce. Today's announcement is further proof that there is no better place to be right now than in the great state of Ohio.[5]

While the governor noted Ohio's talented workforce, he and others in the auto industry realized that transforming supply networks would be challenging. Ideal economic climate or not, producing gasoline engines is fundamentally different than producing batteries. In some ways, the change means that Honda's almost $25 billion investment in Ohio facilities over the past half-century is obsolete.

Luckily for DeWine, Honda, and Ohio, the efforts of Governor Rhodes and many other people in bringing Honda to America in the 1970s have resulted in a well-tuned engine in the form of the Honda–Ohio State University partnership. This partnership is a critical component, a piston, if you will, in the engine that seeks to build an adequate supply network for electrification. Thus, I profile two areas of this Honda–Ohio State University partnership that are critical catalysts in this effort.

The partnership is a five-decade marriage that has benefitted both organizations and is particularly important in the revolution. A description by numbers is impressive, yet I will also tell the stories of two Ohio State faculty members and the teams they lead to "Drive toward the next generation of mobility systems."[6]

The partnership spends more than $100 million annually on research, with over one hundred faculty and staff supervising and teaching over five hundred students. A quick search on LinkedIn reveals almost three thousand Honda people with Ohio State ties. A substantial portion of research funding is provided directly by Honda (and other automotive companies), with another significant portion of funding coming from US government agencies including the Department of Energy and the National Science Foundation. An overview of key inter-disciplinary centers and institutes includes:

- The Center for Automotive Research or CAR
- Center for High Performance Power Electronics
- Driving Simulation Laboratory
- The Simulation and Modeling (SIM) Center

CAR Started It All

Giorgio Rizzoni is an affable Italian born in Bologna who often makes cappuccino for a guest while discussing any and all aspects of automobile design and production. Unsurprisingly, he loved motorcycles and cars and began rebuilding carburetors to extract more power from them in his teens. Enrolling at That University Up North (TTUN—or U of Michigan for non-OSU people), Rizzoni was influenced by an electrical engineering professor saying, "This is circa 1980 . . . and electronics were beginning to find their way into vehicles. So next thing happens, I graduate and I begin working in this lab [the Vehicular Electronics Laboratory led by Prof. William B. Ribbins]—an electrical engineer who was a guest of the automotive lab." Giorgio was in love. He earned a PhD in Electrical and Computer Engineering, writing a dissertation titled "A Dynamic Model for the Internal Combustion Engine" and was recruited to Buckeye Nation, joining the Department of Mechanical Engineering as an assistant professor in 1990. Rizzoni says, "My hiring was strictly related to CAR because of the Transportation Research Endowment Program (TREP) and the TRC-Honda-OSU engineering connection."[7] TREP was created using a portion of the $20 million Honda paid to buy the TRC, with $6 million placed in an endowment to fund and support TREP. Having landed in Columbus just as CAR was being created, Giorgio was surprised one day late in the 1990s when the directorship was open and a friend (Dr. Steve Yurkovich, faculty emeritus OSU) suggested he throw his hat in the ring and coach Rizzoni through his inexperience. After an internal search for the director position, Dean David Ashley then offered him the director position. Fast forward almost a quarter century, and Professor Rizzoni is beloved by many across the industry, and he has been director of CAR since 1999.

As Rizzoni describes it, "The number of CAR students who are working today at Honda or have worked at Honda is uncountable. So, from a human resources perspective, there is certainly a lot of value. Honda has been generous to us in that every time we have asked for support for some of our student projects, you know they're always there ready to assist." The OSU-Honda relationship serves as a conduit for educating and developing talented engineers that move into the industry and Honda to design and build the cars that drive us. Yet current times offer unprecedented challenges in terms of the educational content. Rizzoni continues, "Years ago, the subject areas in which we offered short courses (for Honda associates) ranged from advances in ICE to introduction to ICEs to hybrid electric vehicles." A true scholar and Honda champion, Rizzoni points out that Honda offered the first hybrid—the Insight, beating the Toyota Prius to market in North America by seven months in 1999. Unfortunately for Honda, first-to-market did not lead to a dominant position as hybrids did not reach one hundred thousand sales until April 2005, helped by the addition of the Civic and Accord.[8]

Returning to mid-year 2023, Honda is a nervous giant with expertise in ICE and an inferior position in EV investments and knowledge. It was an uncharacteristically humiliating moment for the auto giant when Honda CEO Toshihiro Mibe admitted on April 25 that the company had fallen behind in the global race to electric vehicles. Mibe outlined his vision to claw back lost ground while providing Honda's annual business briefing. He reported that he and other executives had been shocked at the Shanghai auto show earlier that month. COO Shinji Aoyama described the reaction to a flood of sophisticated, advanced EVs from Chinese brands: "We were overwhelmed by the Chinese." CEO Mibe expanded, "They are ahead of us, even more than expected. We are thinking of ways to fight back. If not, we will lose this competition. We recognized that we are slightly lagging behind and determined to turn the tables."[9]

Are Professor Rizzoni and other OSU faculty and students fighting together with Honda as partners? Yes, they are. Their main focus is on training and developing the workforce in the EV industry. A report titled "Supercharging our Electric Vehicle Workforce" issued in June 2023 by the governor's office and the Ohio Manufacturer's Association has predicted that the EV industry will create 25,400 jobs in Ohio by 2030. Lt. Governor Jon Husted has expressed his support for this initiative. "The advent of EVs necessitates comprehensive changes to our power grid, workforce, and production systems, and Ohio is poised to play a significant role in meeting these demands and accelerating their adoption across the United States."

The report notes that Ohio is the largest producer of automobile parts and is home to the second-largest workforce dedicated to automobile production in the United States, although many in the Hoosier state claim that Indiana is second. Emphasizing both the tremendous opportunities presented by the revolution and the swift pace of change, Ryan Augsburger, president of the Ohio Manufacturing Association, said,

> Today, the automobile industry faces new opportunities and challenges brought by the transition to EVs. Not since the implementation of the assembly line has America's auto industry faced so much change so quickly. Much is at stake for Ohio during this transition.[10]

CAR and Rizzoni are working with Honda by offering three short courses that are a world apart from those offered as recently as five years ago.

These are two-day courses that can be tailored to fit either technical professionals directly involved in the EV development and production pipeline, or offered to other associates who are either not directly tied to vehicle development or are still primarily in the traditional ICE business. In all cases, the Honda-CAR partnership seeks to improve the Willingness to Commit to the revolution. The first course, "Electrification of Mobility," focuses on the drivers and why the industry is moving in this direction, describes the key components, and then identifies challenges and potential benefits. Dr. Rizzoni himself teaches this course. The second is titled "Electric Machines and Power Electronics for Electric Mobility" and is taught by Matilde D'Arpino. Like Rizzoni, she was born in Cassino, Italy, with a love of cars and earned a PhD in electrical engineering. An expert in topics related to power electronics, electric drives and electric traction systems, Professor D'Arpino is an assistant professor, author of more than fifty scientific papers, and a lead principal investigator or co-PI on several research projects funded by the US Department of Energy, NASA, and several major automotive companies. The third short course is taught by Marcello Canova, who—you guessed it—was born in Italy, earned a PhD in mechanical engineering from the University of Parma. The third course is titled the "Energy Storage Systems for Electric Mobility" and covers two days on the fundamentals of Li-Ion batteries.

To this point, I have primarily described the OSU CAR–Honda partnership as a two-party relationship. At the same time there are similar partnerships in other US states and worldwide. CAR and most, if not all, similar centers work with many partners and sponsors. One of the agencies seeking to build capabilities is the US National Science

Foundation, which runs a program titled NSF Engines that was authorized by the US Chips and Science Act of 2022. The NSF Engines program solicits proposals for two types of awards—Type 1 of $1 million and up to two years duration and Type 2 of up to $160 million in funding and a ten-year duration. The NSF Engine program states, "The Administration [Biden] and Congress recognized the value of advancing transdisciplinary, collaborative, use-inspired and translation research and technology development in key technology focus areas.[11]

In May 2023, NSF awarded "the first-ever NSF Engines Development Awards to forty-four unique teams spanning forty-six states and territories and run by businesses, nonprofits, universities, and other organizations. Through these up to $1 million planning awards, NSF is seeding the future for communities to grow their regional economies through research and partnerships. These two-year awards will unleash ideas, talent, pathways, and resources to create vibrant innovation ecosystems across the United States."[12]

The evaluation of research proposals is rigorous and intended to be scientific and unbiased, yet when awards are in the high nine figures and involve partnerships across states and institutions, there can be a definite political element. In the particular case of electric vehicles, there may be difficulties in achieving perfect concordance between US federal agencies such as the Department of Energy and the NSF. There is a need for both basic lab science research in clean energy systems, and for effective methods to bring those systems to market and encourage consumer adoption. Similarly, as emphasized throughout this book, the revolution requires rethinking and redesigning the entire transportation ecosystem. One of the questions is, who should pay for this? Governments? The automakers? As I write this, OSU CAR is part of a large partnership with a Type 2 proposal under serious consideration for funding with the NSF Engine program. The NSF has approved several preliminary review stages of the proposal titled Electrify Mobility Innovation Engine (EMIE), and it is one of multiple finalists competing for final approval. The leadership team for the proposal meets regularly and has contingency plans to execute the bulk of the proposal with private-sector funding (i.e., the automotive manufacturers will fund it themselves) should the NSF not approve the award. Figure 7.2 shows the innovation system assembled by the principal investigators who developed and submitted the NSF grant proposal.

The EMIE proposal lists Dr. Rizzoni as the principal investigator, with four co-PIs, each representing the leading land-grant university in one of the five largest automotive-producing states. These states include Michigan, Ohio, Kentucky, Tennessee, and Indiana, accounting for over 120,000 jobs tied directly to manufacturing. Sixth on the list? California. A comprehensive public-private partnership, the EMIE proposal also includes thirty-one for-profit companies. These companies include automobile manufacturers (Honda, GM, and others), suppliers (Borg Warner, Robert Bosch, and others), logistics providers (FedEx and UPS), battery companies (LG Energy Solution and XS Power), and energy companies (AEP and RevCharger among others). The effort and coordination needed to develop a competitive proposal of this type are enormous. Each of the partner agencies was required to submit a letter of intent to participate. Upon awarding a grant, all partners agreed to establish EMIE as an independent, nonprofit entity to lead the ecosystem. Finally, any major US grant award comes with substantial controls and auditing activities.

So, what does EMIE seek to do? The answer is many, many things EV-related, but here I focus on the educational component. The EMIE Innovation Ecosystem includes community colleges in the five participating states and primary education partners. The revolution requires a

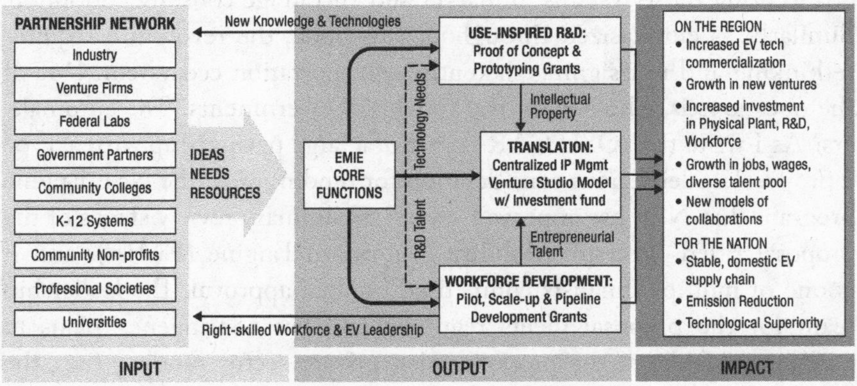

Figure 7.2. The EMIE Innovation Ecosystem

retraining of the workforce at many levels—from the engineers designing the cars to the manufacturing workforce building the cars, to the sales and service people at dealerships and service shops. The combined investment to date in three existing Honda plants (Marysville, East Liberty Assembly, and Anna Engine Plant) is over $8 billion, employing nearly ten thousand associates. The Honda Anna engine plant recently produced its thirty-millionth engine. Given the vast differences between ICE and electric, the anxiety regarding the future of these jobs and associated investment is palpable.

Training, education, and development must be carefully designed and delivered to various groups of people in this ecosystem. Dr. Rizzoni describes the differing educational approaches depending on the audiences:

> But the point is that if an HR person came to any one of our two-day courses, they would be lost immediately because they don't know the second law of thermodynamics or Newton's first law, etcetera. We are designing courses for anyone in the company to acquire and become familiar with the vocabulary of electric mobility.[13]

I have focused on the Ohio State University partnership with Honda to tell a clean story. At the same time, the EMIE grant proposal illustrates the breadth of effort across the industry, particularly in the midwestern heart of automobile assembly in the United States. This entire effort seeks to build Willingness to Commit throughout the manufacturing companies sponsoring EMIE and their supply networks. Education is the key to Transforming the Supply Network.

NEW POWER SYSTEMS MOVE FROM THE LABORATORY TO MASS PRODUCTION

Another Ohio State–Honda partnership involves developing and sharing battery knowledge. In January 2023, the US Department of Energy announced a $3.8 million grant from the Electric Vehicles for American Low-Carbon Living (EVs4ALL) program. This grant is one of twelve funded nationwide to mitigate technology barriers to EV adoption. Anne

Co, professor in the Department of Chemistry and Biochemistry, is the principal investigator for the project. Ohio State collaborators include Professor Marcello Canova and several other CAR researchers. According to Professor Co,

> Collaboration is essential for the creation of innovative technology. It is as exciting as it is rewarding to work alongside colleagues and industry partners working together to find effective and affordable solutions that will advance science and preserve our environment.[14]

The EVs4ALL grant is managed by the DOE's Advanced Research Project Agency-Energy (ARPA-E), a descendant of the original ARPA, which managed the development of the early internet. The goal is to expand domestic EV adoption by developing batteries that last longer, charge faster, perform efficiently in freezing temperatures, and have better overall range retention. Importantly, the team at OSU includes a technology-to-market team led by Professor Jay Sayre, assistant vice president in the Office of Research and the director of innovation for the Institute for Materials Research. As discussed, there is some tension between the private and public sectors. In general, government agencies such as the DOE and NSF seek to support basic or bench science that can be propagated to the world for all to benefit. In some cases, there is a belief that companies in the private sector should be paying for things that help them directly. Regarding the OSU grant from the EVs4ALL, the bulk of the money will be spent on lab science, not translational science. At the same time, a technology-to-market team indicates a strong awareness of the need and value of quickly and efficiently taking discoveries from the lab to the manufacturing plant.

The world has almost a century and a half of experience manufacturing internal combustion engines. It may be fair to characterize the state of art in producing batteries to power vehicles as equivalent to what rolled off Henry Ford's Highland Park plant with Model Ts in the 1920s. Therefore, in addition to the inherent needs to advance the state of art and performance in the batteries themselves, auto manufacturers are desperately seeking ways to speed their learning curves regarding production of these batteries. Common challenges with regard to production include:

- Fulfilling safety requirements
- Safeguarding quality
- Rising costs
- EV batteries require optimization for safety, durability and performance[15]

Briefly considering these in order neatly illustrates production challenges. In terms of safety, many consumers are worried about battery fires, since battery cells must be operated within a specific temperature range. Car designers can protect batteries in a collision with stacks of cells reinforced with lateral bracing, yet this comes at greater complexity and cost for production. In safeguarding quality, there is a trade-off between speed to market and steps taken to ensure a safe, quality product. The reader may recall the $2 billion LG agreed to reimburse GM in 2021 for the recall of over 140,000 Chevy Bolts with defective batteries.

Nobel Prize–winning economist Herb Simon once observed that "Energy and information are two basic currencies of organic and social systems" with a new technology that alters the availability of either working profound changes. As the world scales with EVs, the economics of power and information are in flux. Rivian CEO R. J. Scaringe recently stated that the battery supply chain is mostly nonexistent, that "90 to 95 percent of the (battery) supply chain does not exist." He also added that the current semiconductor chip shortage is just a small preview of what we can expect to experience in the next two decades regarding battery cells.[16]

One of the key partners working with Honda and other automotive manufacturers is Jay Sayre, the director of innovation for the Institute for Materials Research at Ohio State University. Prof. Sayre also leads the team translating knowledge and assets from the Institute to market. In an interview in early January 2023 Prof. Sayre talked me through some fundamentals of how Li-Ion batteries work, his team's research on advancing them, and a planned partnership with Honda. Regarding the state of the art for batteries and performance, Sayre said that current batteries were generation A/B and that scientists worldwide were studying potential generation C/D batteries. He projected that future batteries would be less flammable, safer, and have more extended range as well as faster charging times.

At that time, Prof. Sayre and many partners at OSU, Honda, and other auto manufacturers were hard at work developing plans to develop a launch laboratory. By this, I mean a facility in between a scientific laboratory that does basic research with small volumes and a full-scale production plant with far higher volumes of battery production. Such a launch laboratory would provide a facility to train a workforce in production techniques and produce batteries at a higher output rate, short of full-scale production. Similar launch-type facilities are common in other industries, including companies such as Kellogg's. When Kellogg's wants to try a new flavor of Pop Tart, say avocado strawberry, it will not produce the Pop Tarts in home kitchens. Rather Kellogg's and many other companies utilize flexible launch plants that allow them to make smaller batches of trial products.

In early January 2023, Professor Sayre had just received word from the US Department of Commerce that $4.5 million in funding was being provided for the EV battery launch laboratory. In addition, he and other OSU personnel were working with Honda to finalize a memorandum of understanding for an additional $10 million investment in the launch laboratory. Sayre and his team, including Associate Professor Jung Hyun Kim, were already conducting training with Honda associates on battery production. The training was designed as seven one- to two-hour modules on all aspects of Li-Ion battery production. To be clear, while Honda was the lead sponsor at the time of our interview, the forecast was that many other automotive manufacturers would be both sponsoring and participating as time progressed, much as has occurred with the TRC that Governor Rhodes first championed in the mid-1960s.

In November 2023, Honda agreed to invest $15 million in a twenty-five-thousand-square-foot renovated facility to be housed on Ohio States's campus. The facility aims to accelerate the development of battery cell materials and manufacturing technologies for EVs and provide a learning environment for workforce development. Other partners in this battery transition laboratory include Schaeffler Americas, economic development corporation JobsOhio, and the state government.[17]

In this chapter, I have attempted to provide an unbiased portrayal of Honda's size, history and its collaborations with the government, universities, and corporations. Potential government sponsors include the NSF, DOE, and Department of Commerce. Billions, even trillions of dollars are at stake in this race to revolution. Meanwhile, Honda CEO Mibe attended the official groundbreaking of the $3.5 billion joint venture battery plant in Jeffersonville, Ohio, on February 28, 2023, two months before he admitted in Japan that Honda was behind and trying to "fight back." The leaders of the LG Energy and Honda joint venture, CEO Robert H. Lee and COO Rick Riggle brandished photos with renderings of the planned facility and said,

> It is an honor to represent two great corporations, Honda and LG Energy Solution, both with a long, proud history of success. LG Energy Solution is the leading battery manufacturer globally and is investing aggressively to meet demand for electrification. We are excited to embark on this partnership with Honda, a leader in the global auto industry with a reputation for quality and reliability"

Lee remarked, "If we harness these strengths, I have no doubt our joint venture will be the most successful battery plant in the world, and we look forward to being a part of this massive transformation toward sustainability.[18]

I too am excited to see Honda embark on this partnership with LG Energy; however, at the same time, there are many reasons to be pessimistic. Manufacturing batteries is extremely difficult. Another Japanese company, Sony was part of a recall of over nearly 10 million laptop batteries it manufactured between August 2003 and February 2006. Some of the batteries overheated and caused fires in Dell computers. A spokesman for Sony noted in October 2006 that "the recall was not a safety issue. This is about addressing people's concerns which have become a social problem. And we made a managerial decision that the recall was necessary."[19] At the time, the recall was estimated to cost Sony over $400 million at a time when its vaunted stock value had dropped over 40 percent in the previous five years. To bring this back to Honda, LG Energy and the challenge facing the two, in looking up the battery on my circa 2020 Dell laptop, it is a 0.05 kWh battery. The Honda Prologue, the first EV offering Honda plans, is to come with an 85 kWh battery, nearly two thousand times that for my laptop. That provides two thousand times as many opportunities for the battery to fail. As the saying goes, you do the math—successfully building and launching an EV scale manufacturing plant is an enormous challenge.

Chapter 8

Power Sources

1885: London, New York, and Baku, Azerbaijan

The steam age revolutionized transportation, allowing railroads to open up the western half of North America for settlement, connected Europe and Asia across continents, and facilitated ever faster steamships, dramatically cutting ocean transit times. Titans of business included Rockefeller, Vanderbilt, Rothschild, and Samuel. The last name is less familiar, yet it provides a gateway to the possibilities of power and fuel transitions. The *Sviet*, Russian for "light," sailed up the River Thames into London in 1885. The first oil tanker in history, the *Sviet* was powered by steam created by boiling water, which was generated by burning coal. The seminal feature of the *Sviet* was the two gigantic tanks belowdecks bursting with kerosene from Russia. This was before the first automobile powered by gasoline.

Carl Benz developed the first gasoline-powered car and named it the Benz Patent-Motorwagen. It featured three wire wheels (unlike horse-driven carriages, which were wood), with a four-stroke engine that sat between the rear wheels and two roller chains to the rear axle (Figure 8.1). Patented in January 1886, the Motorwagen was difficult to control and collided with a wall in an early exhibition.[1] Carl sold a few cars, but likely fewer than one thousand, as Henry Ford had not yet developed mass production. Carl's company eventually became the twenty-first-century powerhouse Mercedes-Benz. Product and process innovations often proceed in unpredictable ways, Carl's product was far

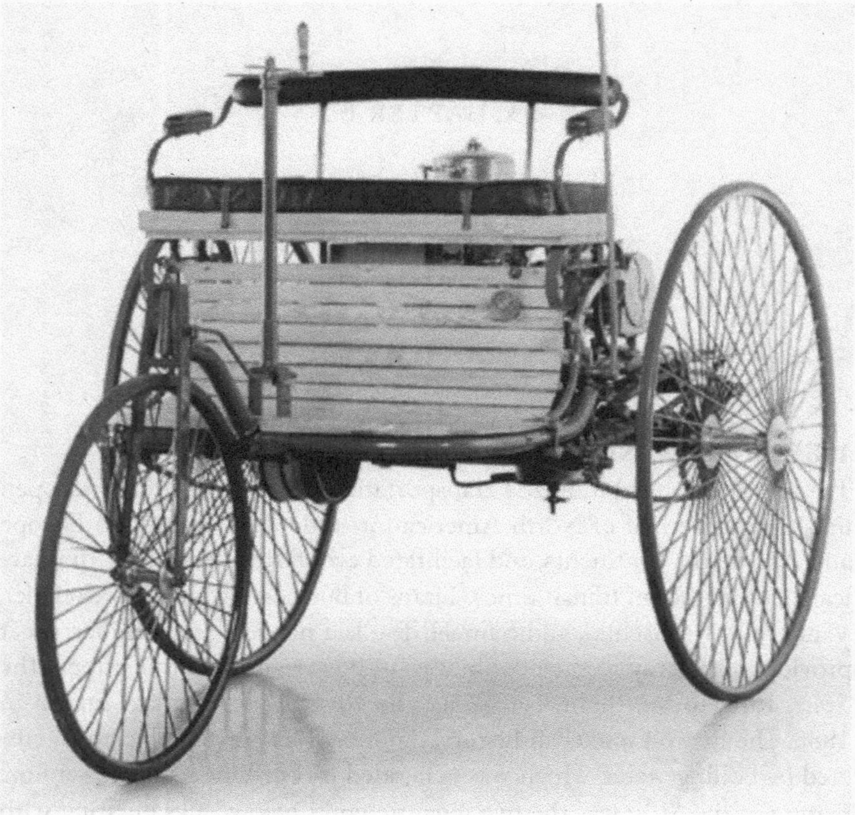

Figure 8.1. The Benz Patent-Motorwagen 1886

ahead of its time since it had to be matched with the mass-production process developed by Henry Ford to be affordable to the masses. A well-off aficionado in 2023 can buy one of the few remaining working Motorwagens for $59,000.

Before gasoline-powered automobiles were developed and widely adopted, the primary use for unrefined petroleum was kerosene. Abraham Lincoln was born into deep poverty in Kentucky in 1809. His rise to the presidency of the United States was fueled in part by many nights reading by the light of a flickering candle. That and whale oil presented the only options for nighttime illumination until the rise of kerosene. In 1858 along the banks of Oil Creek, Edwin Drake drilled the first commercial oil well. Kerosene offered so many improvements on other lighting sources that it singlehandedly killed the American whaling industry. With a peak fleet of 199 whaling ships in 1858, the fleet essentially disappeared throughout the Civil War, falling to 39 ships by 1876. According to the Schumpeterian theory of creative destruction, when an industry threatens another with a better product or service, there will be winners and losers. In this case, the whaling captains of the northeastern United States lost, while John D. Rockefeller and his partners in Standard Oil won.

In September 1916 Rockefeller became the first person ever to reach a nominal personal fortune of $1 billion. By his death in 1937, an estimate of his net worth converted to today's currency values (using his net worth as a percentage of the US GDP at the time) placed his net worth in the range of $300 to $400 billion.[2] The development of an almost complete monopoly was fundamental to Rockefeller's wealth and Standard's success. Not surprisingly, many people had ambitions to enter this lucrative business and compete to win riches.

The kerosene in the holds of the *Sviet* came from the rich oil fields of Baku, Azerbaijan. This was a huge supply network challenge to Standard Oil and Rockefeller's monopoly. Standard had been exporting tins of kerosene from America for over a decade. Notorious for his "cut to kill" strategy, Rockefeller relentlessly reduced prices to squeeze out competitors. Yet the novelty of the *Sviet* lay in two key aspects of the ship itself and the source of its commodity. First, the oil came from a huge field in Baku controlled and owned by Ludwig and Robert Nobel, brothers of Albert, the inventor of dynamite and bestower of the funds for the Nobel Prize. At roughly seventy-seven thousand barrels of oil per day, the Baku fields provided an alternative source to the oil produced by Standard in America. Critically the Baku fields were much closer to Europe, presenting a dire threat to Rockefeller and Standard's monopoly. But first, the oil had to be brought to markets in Europe—and/or Asia.

The *Sviet* was a vast improvement in efficiency—its two voluminous tanks could carry oil at a dramatically lower cost than placing it in tins. The tins were employed to minimize the risk of fire. Fire on ships has always been an imminent threat, particularly in the mid-1800s when ships were built of wood and into the steam age, where steam boilers often blew up or caught on fire. While the *Sviet*'s large tanks were much faster to load and unload oil, thus reducing transport costs substantially, the tanker also had significant vulnerabilities. When large quantities of kerosene were added, things only worsened. The kerosene tended to slosh from side to side in anything but the smoothest seas. Yet more concerning, seamen noticed that combustible petroleum gas tended to collect belowdecks; thus "sailors regarded shipping [serving] on a steamship laden with oil in bulk as nothing less than suicide."[3]

Born in 1853 in Whitechapel, London, Marcus Samuel was of Iraqi Jewish heritage. His father, also Marcus Samuel, ran a successful import-export business selling, among other things, seashells. Much of the early success of M. Samuel & Co., which the son took over in 1878 upon his father's death with his brother Samuel, involved leveraged connections with trading firms in the Far East, namely Japan. By 1889, M. Samuel & Co. earned Marcus and Samuel a comfortable living. Yet Marcus was sensitive to the intricacies of social status, and while the Prince of Wales famously was quite close to the English Rothschilds, Marcus did not enjoy the same status. It mattered little to the English royalty that the Rothschilds were Jewish and immensely wealthy. Marcus Samuel hungered for more, much more.

In 1886 all of the oil produced in Pennsylvania (predominantly controlled by Standard and Rockefeller) amounted to 25 million barrels of crude oil annually. The recently opened Bibi Heybat well in Baku produced seventy thousand barrels daily, roughly 28 million per year. Robert and Ludwig Nobel controlled this well and many others in the region. Additional players included Alphonse and Edmond Rothschild, scions of the fabulously wealthy French House of Rothschild. The Rothschild brothers had come into possession of a railroad that connected Baku, Azerbaijan (the original source of crude oil), and Batumi, Georgia, on the Black Sea. Separated by 560 difficult-to-travel miles through the Caucuses mountains, these two cities represented a way to move black gold from Baku to new markets [particularly in Asia where Standard had no presence] fairly cheaply. If oil could be taken by train from Baku to Batumi and on to Europe and Asia, huge opportunities were available.

Unfortunately for the Nobel and Rothschild brothers, Rockefeller was not one to surrender easily. Following a vicious price war, Standard forced the pairs of brothers to settle, forcing them into a rigged market through its monopoly powers. The various sides engaged in price fixing and unwritten agreements regarding how much oil they could and would sell.[4]

The Rothschild and Nobel quartet of brothers did well, selling thirty-eight thousand tins of oil annually. Yet they were limited by agreements with Standard and the immense danger and unreliability of transporting oil via the *Sviet*. Oil demand was 3 million tins per year in 1890, yet the estimate was that lower-priced oil would expand the market to ten times that amount or 30 million tins per year. Meanwhile, Baku's oil fields spilled up to 1.4 million barrels every two weeks. This oil did not have a market or a storage medium; thus, a giant lake of crude oil grew.

Everyone knew the "solution." A trip of over 18,000 miles by rail and ship was enormously expensive, requiring a 561-mile train ride from Baku to Batumi, then sailing through the Black Sea, the Mediterranean, around the Horn of Africa, and past India to China, Japan, and other emerging markets. Sailing a ship through the Suez Canal would cut the distance in half. Officially opened in 1869, the canal reduced the distance from London to the Arabian Sea by 5,500 miles or approximately ten days' travel time. Alphonse and Edmond's father, Lionel de Rothschild, financed the British government's 1875 purchase of a controlling interest in the Suez. Seems like a simple solution: move the oil through the Suez? Not quite.

To sail through the canal required approval from Lloyd's of London. Fearing a calamity, Lloyds would not let a powder keg like the *Sviet* travel the Suez. Well over a century later, the 1,300-foot container ship *Ever Given* got stuck at the southern mouth of the Suez on March 23, 2021, bottlenecking the travels of at least one hundred ships of similar size on each end of the canal. Lloyd's investors in the 1880s took calculated risks, but they were never going to let the *Sviet* through the Suez.

Standing at the docks in Batumi in 1890, Marcus Samuel saw a solution—an oil tanker, but one far safer than the *Sviet*. On his return to London he met with James Fortescu Flannery and entered into arrangements to design a new type of oil tanker. Intriguingly, Flannery's obituary mentions nothing about the ship he designed for Samuel. Instead, it describes how he was elected to Parliament, serving for over two decades and receiving a knighthood in 1899 and created a Baronet in 1904.[5] To the victors go the spoils, yet in the beginning, the victors must take calculated risks and back them up with engineering, business, and supply chain innovation.

The ship Flannery designed was named the *Murex* (Figure 8.2) and had several special features that were seminal. At 349 feet long and 43 feet wide, it was about one-third the size of the *Ever Given*. Double-bottomed with three separate compartments for stability, the *Murex* had nine transverse bulkheads for strength as well as limiting the damage in case of explosion or fire. The ship could carry four thousand tons of oil, a 135 percent increase on the *Sviet*. Additional features included airtight cofferdams on the oil tanks and two massive pumps to offload and onload oil in twelve hours. Crude oil is extremely sticky and messy, often called black tar, leaving residue and sludge on anything it touches. The *Murex* was also equipped with steampipes to clean the tanks for backhauling—an early form of reverse logistics. Finally, the ventilation

Figure 8.2. The Murex Oil Tanker

system could suck 7,500 cubic feet of air to minimize the chance of a random spark starting a conflagration. Brilliant, and extremely effective, as history would prove. Yet, it was not sufficient.

While Marcus Samuel made a solid decision to hire Flannery to design the ship, another decision involved which firm would construct the ship, in this lay Samuel's brilliance and a bit of political acumen. William Gray & Co., situated in West Hartlepool, England, was founded in 1863 and began building iron-hulled ships. A flourishing business, by 1878, William Gray employed two thousand men and set a British record with eighteen boats launched in a year. In signing with William Gray, Samuel was making an excellent bet on the cost and quality of the ship he envisioned. Of much more importance politically, William Gray was also the regional representative for West Hartlepool on the executive committee of Lloyds of London. Beginning in 1764, Lloyd's published a registry of ships to assist underwriters and merchants in business. Following the British acquisition of the Suez Canal, Lloyd's also determined which ships could transit the canal. Long story short, while there was substantial political maneuvering and quite possibly some financial incentives of a dubious nature, Samuel received clearance to transit the canal with the *Murex*.

The *Murex* and her sisters' design, construction, and successful operation made Samuel a wealthy man and opened up the Far East Market to his partnership with the Rothschilds. It also indirectly led to a knighthood when the *Murex* assisted in freeing a British battleship, the HMS *Victorious*. On Valentine's Day 1898, the battleship ran aground at the mouth of the Suez, well over a century before the much larger *Ever Given* repeated the act. Unable to dislodge the battleship after several days, leaders in the British Navy were pleased yet mortified when the *Murex* was able to dislodge her. Thus, the striving merchant trader was awarded a knighthood, becoming 1st Viscount Bearsted.[6] Samuel's later business moves led him to name his company Shell after his father's original shell-trading shop and eventually partner with Royal Dutch Oil to form Royal Dutch Shell, which today has a value of over $200 billion.

PRESENT DAY

Much like Marcus Samuel, the Nobel Brothers, and Rockefeller, fuel sources and supply chains are rapidly evolving today. Both companies and consumers hope we can transition from dirty, carbon sources of power to clean, renewable sources, including wind, hydropower, and solar. The transition will not be smooth. Just as many people were involved in the journey from tins of oil shipped by sailboat or steam to the *Sviet* to the *Murex*, there will be many actors in the generation of power, storage in batteries, and transmission. Many actors will fail, and some will be fabulously successful, like Viscount Bearsted.

In all cases, there is an interaction between willingness to commit, ability to profit, and transforming supply networks. This chapter examines these challenges and opportunities at present, and I explore the most significant catalyst for change in the automobile industry—power sources, namely electric versus gasoline. To be clear, this is far from the only area in which automakers are innovating to provide greener cars, both in terms of profit and the environment. Chapter 9 will look at some radically innovative efforts with regard to tires. At the same time, without a cleaner power source—electricity or hydrogen [which will be briefly examined]—we are back to dirty gasoline-powered vehicles or horsepower. Thus, this chapter looks at current efforts to move customers and the supply network from gasoline to electric while earning a profit.

Barriers to Adoption

If you visit any social media platform and start browsing commentary on EVs, you will find fanboys, fangirls, and many haters. Gasoline-powered cars are a known quantity; they have a century-plus of innovation, engineering, and development behind them. EVs are an unknown quantity. Thus, one of the primary challenges to adoption is *range anxiety*. Around a third of consumers want a car with a four-hundred-mile range or more, while another third would settle for three hundred miles. Yet the average person drives around forty miles a day. Automobile manufacturers must convince skeptical customers that an EV will meet their needs.

Then there is the refueling challenge, or *charging anxiety*. Drivers today have internalized the process of stopping at a gas station, pulling up to the pump, approaching the head (the embedded computer that controls the pump action, takes payment, etcetera), sliding a credit card in, or walking to the counter to pay cash and then pumping the gas, often while wandering into the store to take care of some biological needs. This process is so ingrained in today's drivers that we take it for granted. In comparison, the prospect of buying an EV that requires forty-five minutes to an hour to charge while "only" providing two hundred to four hundred miles of range is anxiety-inducing.

Auto manufacturers and society can overcome this. As Socrates said two millennia ago, "The secret of change is to focus all of your energy not on fighting the old, but on building the new." Stepping back in time offers some perspective on how individuals and society adapt to new challenges and opportunities and how they overcome anxiety. At the dawn of the twentieth century, primary modes of transport were walking, bicycle riding, and horses. Thousands of people, with Henry Ford as one of the giant influencers, have devoted their energies to the horseless carriage. The number of gasoline stations did not exceed ten thousand until sometime after WWI. Early drivers often had to carry their gasoline or plan in advance for where to stop and often ran out of gasoline. Range anxiety, indeed.

While *gasoline station* is the most common term today, until the 1970s, the more accepted term was *service station* since most owners considered their primary business to be repairs and maintenance. A 1982 article in the *New York Times* paints an interesting picture:

> Until after World War II, gasoline selling was mainly an ancillary activity for the station owner, who made most of his money from repairs and maintenance. The postwar boom in travel began to change that, as soaring demand for gasoline made higher sales volumes possible. With the suburban explosion and the construction of the interstate highway system, the service station peaked in numbers. It also changed its appearance, as the landscaped ranch style stations of the 1950s and 1960's, larger than ever, looking as much like houses as commercial establishments so as to please both suburbanites and their zoning boards, set the tone. . . . The result of all this, according to the API study, was a significant decline in the number of stations. Along with

> the Alamo replica at LaBranch and Franklin Streets, they died literally by the thousands. Various organizations' data differ, but the decline ranged from 22 to 38 percent from 1972 to 1978.[7]

The quote from the *New York Times* is interesting and assigns much of the causation of the decline in gasoline/service stations to the surge in sales volume for automobiles. My argument is that the article accurately assessed the direction yet missed the causation. This was partly because the 1970s and 1980s represented what I labeled an insurgence rather than a revolution. The photo shows a Mobil service station from the 1960s. Figure 8.3 shows the evolution of gas stations in the United States. Two aspects of the graph immediately grab the viewer's attention. First is the rapid growth in stations prior to WWII, the second is the drastic drop in the 1970s. So what explains the severe drop?

A logical but incorrect first guess is associated with the oil crisis of the 1970s. The oil crisis was similar to the Vietnam War in that both represent a crucible of faith for Americans. Readers of a certain age viscerally

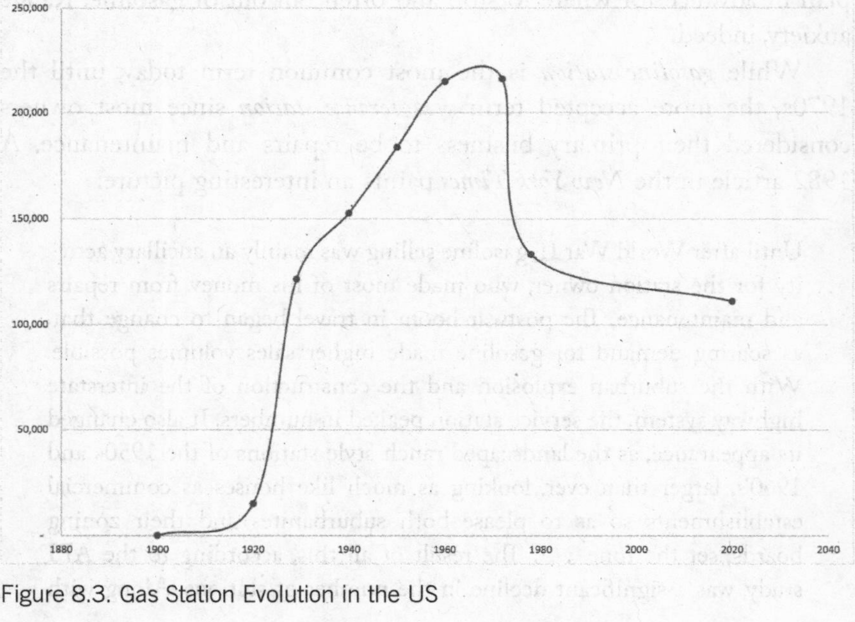

Figure 8.3. Gas Station Evolution in the US

remember the blocks-long lines for gasoline in the 1970s and pulling into a service station with one or more service bays where mechanics would work on their automobile. Perhaps the explanation for the drastic decrease in gas stations is that automobiles became more fuel efficient as Japanese imports, including Honda, Toyota, and Mazda gained market share?[8]

This appears to be a reasonable guess yet is not an explanation for the decrease in gas stations. In the 1970s the United States faced a transition from an economy that relied on domestic oil to one that imported a majority of this key commodity. Many people remember, and some historians employ the term *oil embargo.* A powerful term, yet one that does not adequately peel the layers of the onion to separate cause and causation. US oil production peaked in 1970 concurrent with a dramatic rise in imported oil. In supply chain terms, the US switched sides on the classic "Make or Buy" question that purchasing and supply chain employs. In other words, the US became a net importer of oil, which was very damaging to the American psyche. Further, the universality of oil dependence has shaped world economic and military history for the past five decades. Yet, as psychologically and economically damaging as the oil crisis was, it was not the cause of a decline in gas stations.

The more likely cause underlying the decline involves improvements in the reliability and quality of automobiles. The book *The Machine That Changed the World,* published in 1990, contrasted two fundamentally different business systems—lean versus mass production. Widely adopted worldwide, lean production [aka the Toyota Production System] is a system that has been widely researched and proven to offer manufacturers the ability to reduce cost while simultaneously improving reliability and quality. Hugely important and world-changing, but not a revolution in the automotive industry, more of what I have previously labeled an insurrection. In contrast, the ongoing switch to EVs is a revolution requiring the triple Gears of Change—*Willingness to Commit, Ability to Profit, and a Transformation of the Supply Network.*

As the earlier figures and photographs have illustrated, the automotive supply network has constantly evolved. This includes how automobiles are fueled and serviced, where and how the fuel is obtained, and how the cars are manufactured. Let's examine the challenges and opportunities for batteries and other fuel sources, as well as the supply network. Global leaders have fretted about oil sources and fought wars over it for at least five decades. Ideally, the green transformation of the automobile supply network will avoid the wars. At the same time, business and world leaders are focused intensely on securing sources of lithium, cobalt, and nickel. They are consequently worried about national and corporate access to these rare earth minerals that are essential for the revolution.

From Gasoline to Greener and Better?

The primary emphasis for light vehicles in the automotive industry is a trillion-dollar bet on lithium-ion batteries, with a few bets on hydrogen-based fuels for heavier vehicles. The father of the lithium-ion battery is considered to be John Bannister Goodenough. Born in 1925 in Jena, Germany, to American parents, Dr. Goodenough graduated from Yale University and served as a US military meteorologist in World War II. Dr. Goodenough earned a PhD in physics at the University of Chicago, where one of his advisors was Enrico Fermi, who has been called the "architect of the nuclear age." Fermi built the first nuclear reactor underneath the University of Chicago football field. Fermi passed away prematurely at fifty-three following a diagnosis of inoperable stomach cancer in October 1954. He suspected that proximity to the nuclear pile he developed was a leading cause of the cancer. Two of his graduate assistants who also worked near the pile also died of cancer.

Luckily for the world and Dr. Goodenough, Fermi's student, lived a much longer life. He spent twenty-four years at MIT's Lincoln Laboratory before moving to the University of Oxford in the late 1970s. In England, Goodenough built on prior work by Stanley Whittingham, who discovered in 1980 that LixCoO2 could be used as a lightweight, high-energy density cathode material. This allowed a doubling of the capacity of lithium-ion batteries. Commercialized through Sony by Akira Yoshino after further refinements, Goodenough's research resulted in his receiving the Japan Prize in 2001. In 2019, Whittingham, Yoshino, and Goodenough were jointly awarded the 2019 Nobel Prize in Chemistry along with Whittingham and Yoshino. Goodenough was the oldest living Nobel Laureate before his passing on June 25, 2023. In addition to the Nobel Prize, Goodenough also won the Enrico Fermi Award in 2009, well over half a century after his mentor's untimely death.

The DNA of a lithium-ion battery begins with a single lithium-ion cell, which consists of four main components: two electrodes, a negatively charged anode and a positively charged cathode. The electrodes are charged by moving lithium ions from the cathode through a separator to the anode. The flow is reversed during discharge. Anodes and cathodes can be produced from multiple materials, including graphite, lithium cobalt oxide, lithium iron phosphate, and lithium manganese oxide. Each material offers different benefits and voltages. Inside a battery, the electrolyte helps transport positive lithium-ions between the electrodes, with the most common electrolyte being lithium salt. The final component is the separator, a thin sheet of material that allows lithium ions to pass through but doesn't conduct electricity. The separator is a critical safety component—if it gets too hot, the pores close and prevent the lithium ions from passing through. In the best case, the battery shuts down; in the worst case, the battery catches fire.[9]

Individual cells can be connected in series or parallel to build a more powerful battery. Tesla's first high-volume EVs, the Model S and Model X had cells known as "18 650" (18 mm in diameter and 65 mm long), while the Model 3 has a larger configuration of cells labeled "21 70" (21 mm in diameter and 70 mm long). The standard or base Model 3 battery pack comprises 2,976 cells in groups of 31 cells per brick. Different battery architectures offer other pros and cons, with the Model 3 battery offering a much less expensive battery than earlier models, yet the battery still was estimated to cost $14,000.[10] Tesla showed proof of concept; now other auto manufacturers are seeking to differentiate their offerings and develop competitive advantages.

In the summer of 2020, GM announced a new battery architecture named Ultium.[11] Mary Barra discussed the new platform: "Our team accepted the challenge to transform product development at GM and position our company for an all-electric future."[12] That all-electric future relies on a collaboration with LG Chem, a South Korean chemical company with the tenth-largest revenues in the world. The company has eight factories in South Korea and a network of twenty-nine business locations in fifteen countries. With worldwide revenues, LG Chem was well positioned to pursue new markets in 1999 when it began Korea's first mass production of lithium-ion batteries. Capitalizing on a grant of approximately $150 million from the US Department of Energy, the company formed LG Chem Michigan and built a production plant for advanced battery cells for EVs, starting production in 2013. The plant can produce sufficient cells to supply between fifty thousand and two hundred thousand battery packs to General Motors and Ford, among others.[13]

GM's strategy is to keep battery cells at a cost of under $100/kWh via its partnership with LG Chem, which, as described in the earlier chapter focused on GM, has had both wins and losses. In contrast, Tesla's primary power source partnership is with Panasonic. Tesla's batteries have decreased from $230 to $127/kWh over the past few years. Putting that in perspective, a Tesla Model 3 entry-level carries a 60 kWh battery, with a purported range of 305 miles per charge. At $230/kWh, the battery costs Tesla over $16,000 to produce, while at $127/kWh a little less than $9,000.[14] While the cost of the battery is a critical concern to potential consumers as it gets built into the price of the car, other significant factors come into consideration as auto manufacturers seek to transform their supply networks.

For industry stalwart GM, breadth of use is of critical importance. At the announcement of the Ultium platform (also often called BEV3) on March 4, 2020, Mary Barra claimed, "What we have done is build a multi-brand, multi-segment EV strategy with economies of scale that rival our full-size truck business with much less complexity and even more flexibility." That flexibility is a critical foundational element of success for global operator GM, as Table 8.1. shows. The Ultium/BEV3 architecture is projected for use in ten different EV models to be produced in at least six other assembly plants in three countries.

Carbon and Financial Characteristics of Li-Ion Batteries

From a consumer perspective, there are many positive features of Li-ion batteries, while there are also many off-putting challenges. Across the auto industry, a critical challenge is persuading customers that

Table 8.1. General Motors Ultium Production Plans

Vehicle	Introduction Model Year	Assembly Facility 1	Assembly Facility 2
Cadillac Lyriq	2022	Spring Hill, Tennessee	Jinqiao Cadillac, Shanghai, China
Cadillac Celestiq	2023	Warren Tech Center, Michigan	TBD
Chevrolet Equinox EV	2023	Ramos Arizpe, Mexico	TBD
Chevrolet Blazer EV	2023	Ramos Arizpe, Mexico	
Cruise Origin	2023	GM Detroit-Hamtramck	TBD
Cadillac Optiq	2024	Ramos Arizpe, Mexico	Jinqiao Cadillac, Shanghai, China
Cadillac Symboliq	2025	Spring Hill, Tennessee	Jinqiao Cadillac, Shanghai, China
Buick Electra E4		TBD	
GMC Terrain EV		TBD	
GMC Acadia EV		TBD	

SOURCE: MARCH 4, 2020, HTTPS://GMAUTHORITY.COM/BLOG/2020/03/NEW-BEV3-GM-EV-PLATFORM-UNVEILED.

electric-powered cars are at least on par with, if not better performing than ICE cars. Generally, electric power is cheaper than gasoline, yet it depends on the source of the electricity as prices differ substantially across the United States and the world. As an example, consider data from my Tesla Model 3. In the past year, it has used 3,294 kWh of electricity. This electricity came from four sources: 68 percent of my charging occurred at home on a slow Level 1 charger. In other words, I plug into an outlet in my garage when I get home, and my Tesla charges at 3 to 5 mph. My house has a detached garage, so installing a faster Level 2 outlet would cost several thousand dollars, which my wife and I have not done. The remaining charging consisted of 10 percent on Tesla's proprietary supercharger network and 4 percent at work on a Level 2 charger that my university installed. As a quick side note, I originally bought a hybrid Toyota Camry in 2012, then bought my current Tesla in part because my university provides special parking spots for low-emission fuel-efficient (LEFE) vehicles [parking at my university is a nightmare experience!]. Finally, 18 percent of my charging needs were at other Level 2 chargers. Tesla estimates that my transportation cost me $386 in electricity and saved me $1,065 in gasoline. My driving in the past year was approximately twelve thousand miles.

Offsetting challenges to this reduction in cost include the need to charge, range anxiety, and mixed outcomes in terms of carbon reduction. Somewhat surprising to my family and me, is that *not once in four years have I run out of charge*, although I did scare myself once or twice. Charging has become relatively routine and habitual, requiring a minute or two to plug in when reaching a destination and unplug when leaving. However, digging deeper unmasks extended challenges. First, it is important to address a key question: Is all electricity greener or less carbon-intensive for the environment? No, it varies greatly by state within the United States and by country around the world.

Let's start by comparing national averages with two contrasting states, California and Wyoming.[15] The national average for renewable power sources (wind, hydro, biomass, geothermal, and nuclear) is 40 percent of all electricity generated. In comparison, the remaining 60 percent comes from nonrenewable sources (most often fossil fuels). California is generally seen as a very progressive and liberal state, with a lot of sun, wind, and water that can be utilized. Thus, over half of California's electrical power comes from renewable sources, whereas Wyoming only gets a little over 20 percent. This raises the question, how "green" from a carbon emissions standpoint are EVs?

The US DOE also provides a calculator that allows users to choose their state to see the annual carbon emissions by vehicle type. Starting with traditional ICE vehicles powered by gasoline, an average driver emits a little over six tons (twelve thousand pounds) of CO_2 annually. Hybrid vehicles lead to improvements, with plug-in hybrids and all-electric cars representing up to an 83 percent decrease from base emissions. There is a dramatic difference between driving an EV in Wyoming and California, and we have not considered the cost or convenience of charging yet.

Combining both cost and carbon emissions yields some stark contrasts for a purely electric vehicle. California, the "green" state, comes in with relatively low emissions—an EV will account for about 1,500 pounds of carbon emissions per year while costing about 20 cents per kWh. This means a 75 perecent reduction in emissions from an ICE vehicle along with a cost savings. In comparison, Wyoming and Idaho are both extremely affordable with costs of less than 8.5 cents per kWh, but Wyoming has *much* higher carbon emissions with 5,700 pounds per year—the EV will be only a slight improvement over an ICE vehicle. On the other hand, Hawaii has the highest US electricity costs at 30.31 cents per kWh as well as very high carbon emissions of almost 4,500 pounds per year, only a 25 percent improvement over an ICE vehicle.[16] At this price, driving my Tesla in Hawaii would cost me around $1,195 per year—still a savings but much less of one. While fueling an EV is almost always cheaper than an ICE vehicle, the relative reduction in carbon emissions depends greatly where and how the electricity is generated.

Why is Hawaii still so "gray"? Simple, 80 percent of its power comes from burning petroleum, which requires *a supertanker delivery every ten days*. Prone to supply chain disruptions? Yes. Expensive? Yes. Carbon emitting? Yes. The good news is that efforts are underway to bring more solar power and storage to the Aloha State.

THE BATTERY SUPPLY NETWORK

Figure 8.4 provides a rough overview of the Li-Ion battery network. While this diagram is relatively complex, it is a fairly high-level representation of several of the critical challenges and opportunities. Beginning with the original power source in the upper right, four categories are identified: wind, solar, and hydro all represent renewable—if not 100 percent—steady sources of power, while fossil fuels, including coal, petroleum, and liquefied natural gas are all fossil-based. Nuclear power represents a stable (maybe renewable) source of power. Still, it is plagued by very high costs and often substantial worry over its safety in both the short and long term. Thus this book focuses on the quartet of renewables plus fossil.

Original Power

Wind power is an attractive renewable energy source, in theory, plentiful and relatively easy to capture, presuming a reliably windy environment. The state of New York has a substantial coastline, over 20 million residents, and huge power needs, making it a logical fit when first-term,

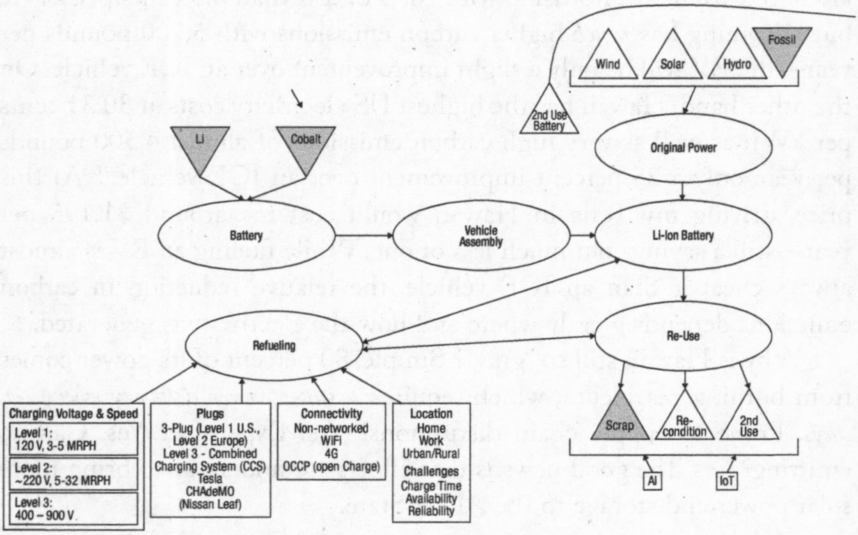

Figure 8.4. The Power and Energy Storage Supply Network

first-ever female Governor Kathy Hochul laid out her plan for developing offshore wind power in the Empire State in early 2022. Hochul promised thousands of jobs and a drastic reduction in carbon emissions. Yet the supply chain and infrastructure to build and deploy wind power in New York didn't exist. Instead, northeast of New York, the smallest state, the State of Rhode Island and Providence Plantations leaped to the early lead in contracting. This is based on a first-mover advantage, since the first offshore wind farm in the United States was finished and connected to the electrical grid in 2015. Sixteen miles off the Rhode Island coast, the Block Island Wind Farm consists of five turbines producing six megawatts each. This thirty-megawatt production capacity is enough to power roughly fifteen thousand homes. Importantly, the Block Island Wind Farm has weathered several significant category 3 storms as it was designed to do without damage and has functioned close to exactly as intended.

Governor Hochul has called for the ability to produce 9 MW of power off the shores of New York by 2035, while a group of environmental activists have pushed for 15 by 2040 and 20 MW by 2050. The manufacturing work in Rhode Island is in support of the South Fork Wind project, expected to be the first to supply offshore wind directly to New York. Thirty-five miles off the tip of Long Island, South Fork is expected to begin contributing to the power grid by the end of 2023. The 132 MW it will ultimately produce will travel through sixty miles of cables underwater to a substation in East Hampton. This is enough to power approximately seventy thousand homes at peak production.

In January 2023, the New York State Energy Research and Development Authority accepted bids for building another wind farm off the Long Island coast. Yet, at press time for this book, the bulk of work for these projects was outside New York. About 150 miles up the coast from New York City, Chris Petit is the shipyard superintendent for Blount Boats in Warren, RI. He manages a crew of almost fifty that welds together aluminum parts to form a ninety-nine-foot catamaran that will carry workers to the South Fork turbines. A joint venture between Orsted, a Danish company, and Eversource, a large New England Utility, the operation is not conceptually much different than Marcus Samuel and his partnership with James Flannery Fortescu in building the *Murex*. In late January 2023 Orsted's operations on the bayfront in Providence included workers sludging through concrete to form a circular platform designed to guard one of South Fork's twelve turbines. Nearby another set of workers assembled internal platforms that will be the mount for the turbines.

Meanwhile, about fifty miles west in New London, Connecticut, across from an aging submarine plant, components for the South Fork turbines, including three-hundred-foot blades will be delivered via ship to State Pier. That aging submarine plant was founded in 1899 and named Electric Boat Division after its primary source of power. WWII submarines ran on gasoline thus had to stay close to the surface and use a snorkel to get air for their ICE. They could only completely submerge on *electric batteries*, which were extremely weak compared to modern ones. The limited power of the batteries meant limited speed, which is what made the submarines "sitting ducks" for depth charges, huge losses during the war and many, many tense submarine chase scenes in movies. Now named General Dynamics Electric Boat Division (EB), the company has produced nuclear submarines for the US Navy since the 1950s. Nuclear power does not need oxygen, thus removing the need for a snorkel on the submarine, the diesel engine and the old, antiquated batteries.

When the Cold War ended in the late 1980s, times changed for the company as orders from the Navy fell drastically. The company completed and delivered forty-nine nuclear submarines during the 1980s and 1990s, only four in the first decade of the twenty-first century and only seven between 2010 and 2020. In short, EB has been in benign decline due to the drastic changes in the needs of the US Navy. Although not directly related to EB, the current opportunity for Groton/New London, Connecticut lies in a $255 million project that involves 150 workers striving to develop a site suitable for the final assembly of turbines for the South Fork Wind project. These turbines will then be loaded onto barges and transported out to sea for installation. While New York is racing to catch up to its diminutive neighbor Rhode Island and sister state Connecticut, it is certainly not alone. Other states announcing major wind energy goals and investments include New Jersey where in late 2022, Governor

Phil Murphy doubled the state's target for offshore wind power to eleven gigawatts by 2040. In addition to New York, New Jersey, and Rhode Island, there are wind projects under development in an additional at least eleven states.[17] Ideally, for residents of sponsoring states, the production work would occur primarily in those states. For example, factories to produce steel support tubes for turbines are being built at a port in Paulsboro, New Jersey, according to David Hardy, the chief executive of Orsted Americas, whose headquarters are split between Boston and Providence. "Those are the long-term jobs, they're thirty-, thirty-five-year jobs," Mr. Hardy said.

Yet until domestic facilities like those are up and running, many of the largest components of the first commercial wind farms in America will come from overseas or smaller states. For now, many smaller states and countries benefit from the changing winds on power. The very long lead times for major construction projects of this type present another high hurdle for the transition to wind. The total pipeline of projects includes an estimated 40 GWh, enough to power over 21 million homes. Yet relatively little of this is operating or under construction, with the bulk being in the permitting, site control, and planning stages. In other words, the large bulk of this capacity will not be available until 2030 or later.

The second huge challenge for original power is the inherent variability of the source, while hydro and wind are relatively stable, they are not completely stable or predictable. The third renewable, solar, has two obvious limitations. First, the sun does not shine at night; second, regions of the world vary substantially in the amount of daily sunshine they receive. The first problem is the more substantial of the two since in most communities peak power demand is the first couple of hours after homeowners come home for the evening when they typically cook dinner, do laundry, turn on all the lights, and jump onto their various electronic entertainment devices—precisely when the sun is setting and the solar power is disappearing. Two approaches are in consideration and experimentation to solve or mitigate this problem. The first approach utilizes differential pricing and internet-connected technology to nudge users to adapt their power usage to match available generation better.

While there is a considerable amount of activity in that area, my primary focus will be on using batteries for storage.

Closing the Circular Supply Network for Batteries—Reuse

There are at least two stages to EV battery life post the initial usage in a new vehicle. Second life is when the battery is taken from the vehicle and redeployed as power storage in an application in which it is not the primary power source for a vehicle. Numerous organizations are experimenting with ways to use second-life batteries. Many organizations are also developing processes to harvest the most valuable components, including Lithium, Cobalt, Nickel, and Copper, from batteries when they are no longer providing acceptable performance in cars or second-life applications. Such third-life applications are intuitively appealing yet the supply, mechanical, and chemical challenges of reusing materials in a third-life application present many challenges, including cost. As in other sectors, creating a truly circular supply chain that captures, harvests, and repurposes materials has numerous hurdles.

In the shorter term, many see second-life applications to store excess power for wind and solar as a doubly green endeavor—green in terms of profitability and in terms of high ability to reduce carbon emissions. With an estimated 10 million EVs worldwide at present, this is projected to be a valuable market, with roughly 1.7 million Li-ion batteries projected to be available by 2030 representing a combined value of roughly $5 billion.

Li-ion batteries degrade as they cycle through power charging and driving cycles. Drivers can expect to travel 100,000 miles before a battery loses 20 percent or more of its total capacity. This means that a battery that once had a maximum range of 250 miles now will only cover 200. The owner can live with this, install a new battery at the cost of 25 to 40 percent of the original car, or buy another car. As a second-life battery for a grid, the batteries remain helpful until the initial capacity drops by 40 percent—another ten to fifteen years of service. The used battery can be sold for small-scale applications. One example is a soccer stadium in Amsterdam, opened in 2018; that employs 150 batteries from the Nissan Leaf, profiled as the first mass-produced EV in chapter 2. The Johan

Cruijff Arena, home of the Dutch football club Ajax has 4,200 solar panels on the roof of the stadium. The second-life batteries from Nissan Leaf cars can store 3 MW of power, enough to charge half a million iPhones.[18] This combined with the solar panels on the roof of the stadium provides a circular, reliable use of clean renewable energy. With the longest history of EV sales, beginning in 2010, Nissan Energy Services believes it has a valuable head start in developing the market and applications. Director Soufiane El Khomri states, "We expect future opportunities to increase as the EV market continues to flourish."[19]

The International Energy Agency estimates that global investment in grid-scale batteries reached $6.8 billion in 2021, up from $4 billion a year earlier. Another use is to capture the raw materials of the battery as inputs to new ones. Companies seeking to capitalize in this area include Redwood Materials Inc. of Carson City, Nevada, and Li-Cycle Holding Corp. of Toronto. Redwood collects and recycles components from Panasonic Holdings Corp., Tesla's main battery supplier. Tesla cofounder J. B. Straubel, who spent fifteen years as chief technical officer at Tesla, announced after leaving Tesla that he was focusing on his startup Redwood Materials in 2019. Just before Christmas 2022, the company held an opening ceremony for a planned investment of $3.5 billion for a recycling facility in Berkeley County, South Carolina. Projected to employ 1,500 people, Redwood also announced a deal with Toyota in June 2022 to recycle batteries from hybrids, including the Prius. In making these announcements, Straubel noted that the battery supply chain would likely be the bottleneck in the industry's transition to EV.[20] Reflecting on his time starting up Tesla, he noted, "People underestimate how difficult it is to ramp a high-volume manufacturing company. . . . I think we will see more pain amongst the field of EV startups. At the same time, there is no question the EV movement is beyond a point of no return."

A substantial opportunity/impediment in achieving a smooth, circular economy for EV batteries and materials is the need to combine basic research on powertrains and technology to monitor usage patterns. In these areas there are numerous private-public partnerships, one of which is spearheaded by Ohio State University Professor Giorgio Rizzoni, who directs the CAR—Center for Automotive Research. As presented in the earlier chapter on Honda, partnerships between public and private entities are critical to a successful energy transition. Collaborative product development is intended to support the entire US automobile industry and help all players advance to gain ground on the Chinese industry.

As profiled in chapter 2, batteries may be *mechanically* simpler with fewer moving parts than an ICE car, however, in terms of *computing needs and wiring*, they are considerably more challenging with the state of knowledge being in its infancy. As illustrated in Figure 8.5, the electric motor is one of the few moving parts. Clearly, choosing a single or double motor impacts a car's cost, performance, and maintenance needs. At the same time, power electronics, thermal management, and other components require extensive computing power and algorithms. As a simple explanation, power is routed to the individual cells in the battery module to balance the lifetime expectation for usage with the speed of each recharge cycle. Consider this to be like your cell phone's message saying it is "optimally charging"—yes you could charge it in one cycle faster, but this would come with a trade-off to the battery's lifespan.

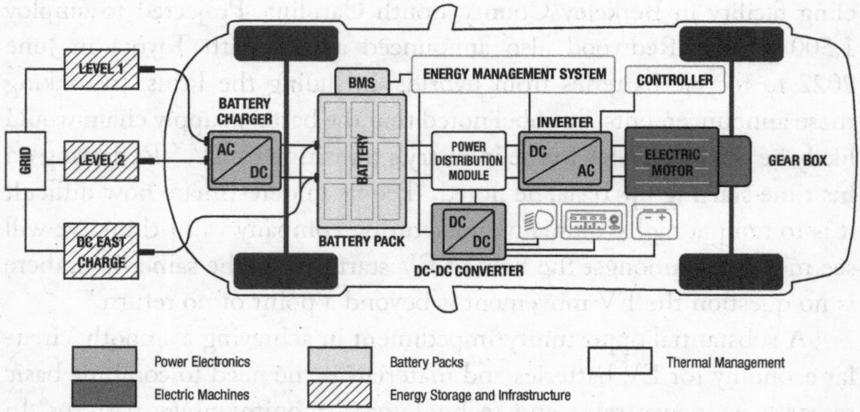

Figure 8.5. Electric Powertrain Components and Systems Requiring Critical Solutions

COURTESY OF OHIO STATE UNIVERSITY, CENTER FOR AUTOMOTIVE RESEARCH, HTTPS://CAR.OSU.EDU/

Artificial Intelligence and Internet of Things in the Li-Ion Supply Network

Returning to Figure 8.4, in two areas of the supply network—manufacturing and reuse of batteries, note two considerations: AI and the Internet of Things (IOT). These technologies potentially combine smart sensors or IOT to collect data on battery usage and machine learning or artificial intelligence to improve the manufacturing process on first use, then to analyze the collected data to stretch battery life during use and pin a value to a used battery. Essentially, AI and IOT have potential to identify which batteries have been overused—charged beyond recommended levels or gone through too many charge cycles to be reliable.

Upstream battery production has a high waste process, with 8–10 percent of the rare materials, including lithium, cobalt, and nickel, being scrapped.[21] There are thousands of variables that affect the manufacturing process. These include mixing (temperature, time, and atmosphere), coating (film thickness and coating speed), drying (temperature, speed, and foil pre-tensioning), slitting (tool wear), calendaring (process settings), and vacuum drying (processing time).[22] AI and IOT sensors are being employed to seek improvements in both the efficiency of converting rare-earth minerals to practical batteries and in reducing the cost to manufacture the batteries.

Downstream, the millions of EVs expected to be produced and sold in the next decade will eventually result in opportunities to recycle these rare-earth materials. Players in this new and growing industry include Cirba Solutions, which broke ground on a $250 million facility to process and recycle batteries in Lancaster, Ohio, with US Department of Energy Secretary Jennifer Granholm in attendance on August 7, 2023. A second key player is Li-Cycle, which is a leading global lithium-ion battery resource recovery company and North America's largest pure-play lithium-ion battery recycler, with a rapidly growing presence across Europe. A third player in this space is J. B. Straubel's Redwood Materials, outlined earlier. Currently, these companies and several others are investing to build out their capabilities to capture and recycle EV batteries.

I will not examine the process of recapturing the materials in a Li-ion battery in detail, but I want to point out that there are numerous competing methods. In addition, the recycling process is complicated because every auto manufacturer is producing fundamentally different batteries—with the primary goals of reducing the cost and improving performance to sell a car, downstream recycling is at best a minor consideration. This means that by the time there are large volumes of batteries that can be recycled it will be challenging to handle the high variety of battery types. AI and IOT will be necessary to make this eco-system function and capture higher amounts of material for re-use.

The use of AI will allow companies to review huge amounts of data generated from the daily driving of EVs. The analysis of this data will allow a choice between sending a battery for a second-use application such as the Johan Cruijff Arena described earlier, sending the battery for recycling, or scrapping the battery entirely. AI and IOT will be critical in this effort as tracking materials to create a circular supply chain is enormously difficult. For example, a recent Greenpeace report estimated that only about 5 percent of first-use plastic gets recycled worldwide.[23] The good news for Li-Ion batteries? They are much more valuable than single-use plastics; thus, there is an opportunity to make this work, although it will be enormously challenging. According to the European Commission, the European Union currently imports about eight hundred thousand tons of car batteries, 190,000 tons of industrial batteries and 160,000 tons of consumer batteries annually. A large part of this is not recycled but scrapped. This is now set to change completely. The new recycling targets for recovery of materials from used batteries will be as follows: 50 percent by 2027 and 80 percent by 2031 for lithium, while the figures are 90 percent by 2027 and 95 percent by 2031 for cobalt, copper, lead, and nickel. Definitely stretch goals!

In July 2023, the logistics giant DHL announced the opening of its EV Center of Excellence in Mexico. In addition to "traditional" logistics expertise, the Germany-based giant seeks to add additional capabilities necessary for the revolution. These include extensive use of sensors and IOT to be integrated with digital solutions for traceability. In addition, DHL staffs the Center of Excellence with specialist teams trained to move hazardous goods and EV batteries combined with complete temperature and humidity monitoring throughout the facility. At the time of writing, DHL was planning to open similar facilities near Detroit, in Europe, and in China to support the growing EV supply network.

A SHORT EXPLORATION OF HYDROGEN AS A FUEL SOURCE

Hydrogen was actually the original fuel source for the first internal combustion engines over two centuries ago. Demand for hydrogen has tripled over the past fifty years and the production of hydrogen is almost entirely supplied by fossil fuels, accounting for 6 percent of global natural gas consumption and 2 percent of global coal use. The production of hydrogen accounts for 830 million tons of carbon dioxide annually, roughly the same as emissions from 140 million ICE cars. Global hydrogen demand reached 95 million tons in 2022 with its primary use being in the refining and chemical sectors.[24] So why am I writing about the potential use of hydrogen for transportation?

The answer is twofold. First, if hydrogen can be produced cleanly then it provides a much more potent power source than electricity. Second, the use of Li-ion batteries for transporting heavy goods—large trucks, trains, and ships—is simply a nonstarter. A multitude of researchers and engineers worldwide are working on methods of producing cleaner hydrogen. The next chapter on tires examines Monolith, a company that is pioneering a new production process for both carbon black, which makes up 30 percent of the weight of every automobile tire on the planet, and hydrogen, which in the case of Monolith is planned to be fed to a partner power plant to produce clean electricity to power the production of carbon black. Hydrogen can be burned in a combustion engine that is similar to today's petroleum powered engines, thus if clean hydrogen can be produced it offers an excellent fuel source for trains and ships.

For automobiles, the most promising application for hydrogen is in fuel cells. As in many things, General Motors pioneered the first hydrogen powered passenger vehicle in 1966, an Electrovan that had a range of 150 miles, could hit 70 mph and accelerated from 0 to 60 in thirty seconds. The biggest barrier to adoption? Then as now a lack of available infrastructure for refueling.[25] How does a hydrogen fuel cell work? It uses compressed hydrogen to convert chemical energy into electricity. The battery is similar to those in EVs with an anode and cathode. When compressed hydrogen is fed into the fuel stack, the hydrogen is broken into positive and negative charges at the anode. The electrical current generated is used to power the vehicle. When the reaction is done the molecules rejoin with oxygen to form water, the only by-product emitted.[26] In August 2023, Toyota debuted a gasoline-electric hybrid version if its iconic Land Cruiser, which in many parts of the world has better brand recognition than the Toyota brand as a whole. *Automotive News* wrote of the new model, to go on sale in mid-2024, "the long-running nameplate is big, boxy, brawny and popular. The question is how to keep the hulking SUV mean and green through the end of the decade in the age of carbon reduction.[27]" Many in the auto industry have argued that Toyota is behind in the race to electrification, while the company's leaders argue they are following a "multi-pathway" approach to carbon dioxide reduction, with both hydrogen fuel cells and solid-state batteries in the mix to supplement and/or replace traditional gasoline as power sources. Simon Humphries, Toyota's global design boss and chief branding officer noted at the hybrid Land Cruiser launch that the company "is committed to providing mobility to everyone in the world, but not everyone in the same situation. Whether it will be a BEV in the future or a hydrogen-powered engine or a fuel cell, who knows what is best for Land Cruiser. There is lots of discussion we have to have."[28]

Certainly there are many hurdles to the widespread adoption of hydrogen as a fuel source, as there are for the widespread adoption of EVs. I have not delved into the challenges of building out the charging infrastructure or the upstream challenges of mining rare earth minerals including lithium, nickel, and cobalt. These all present significant challenges that need to be overcome. The reason for not examining these in depth is simply that my goal was to keep this book more concise and digestible. A recent *New York Times* article on the energy transition "The Clean Energy Future Is Arriving Faster Than You Think," points to work being done, ironically enough at University of Tulsa's School of Petroleum Engineering on hydrogen as a clean energy source.[29] As the saying goes, where there's a will, there's a way.

In closing this chapter, I end with another historical story on Thomas Midgley as an example of the simultaneous brilliance and folly of mankind. Midgely was a brilliant chemist who in a six-year stretch from 1922 to 1928 invented two things that radically changed the world. His second invention was freon, which enabled modern air conditioning and eventually led to a hole in the ozone. The world was able to take action, with twenty-four nations signing the Montreal Protocol on Substances that Deplete the Ozone Layer in 1987. Today, the ozone has largely recovered from near depletion. Thus, the focus here is on Midgley's other huge discovery—unleaded gas.

Prior to Thomas Midgeley, automobiles were plagued by a terrible engine knock, described by Sharon McGrayne in *Prometheans in the Lab* as when "driving up a hill made valves rattle, cylinder heads knock, the gearbox vibrate and the engine suddenly lose power." Midgley set out to investigate the phenomenon by devising a miniature camera to shoot footage inside the piston cylinders. The film revealed that the gasoline was igniting too abruptly inside the cylinders, creating a surge of pressure, unpleasant vibrations and wasting energy. Midgely and his team next set out to find a solution, ultimately testing over thirty thousand different chemical compounds over a five-year time span. The final solution? One part tetraethyl lead to 1,300 parts gasoline—roughly a spoonful of lead solved the knocking problem. Charles Kettering and Midgley were appointed to lead the joint venture company Ethyl Corp. formed in partnership between GM, DuPont, and Standard Oil. In many ways, it was the addition of lead to gasoline that helped automobile sales truly take off. During the 1920s the number of automobiles in the United States tripled, and by 1930 Americans owned close to 80 percent of the cars in the world.[30] By 1935 Ethyl was included in 90 percent of the gasoline sold in America.

As Midgely and Kettering appeared to know at the time of their discovery of Ethyl, it has some very ugly unintended effects. Midgley himself suffered lead poisoning from his work to develop Ethyl, declining to attend a gathering of the American Chemical Society and writing, "I find that my lungs have been affected and that it is necessary to drop all work and get a large supply of fresh air." As the public and scientists gradually accumulated knowledge regarding the dangers of lead poisoning, the US Congress passed the Clean Air Act in 1963 effectively ending the use of leaded gas in new automobiles. Over the course of the 1970s ethanol produced from corn gradually took the place of Ethyl as a fuel additive.

The story of Midgely and leaded gasoline illustrates the great ingenuity of mankind while also illustrating that rarely does a single innovation get everything right. As the automobile and transportation industries search for alternatives to ICE vehicles, we are likely to see numerous stories of mixed success and failure play out over the next decade or more.

Chapter 9

Tires, Tires Everywhere

Carl Friedrich Benz partnered with Max Rose and Friedrick Willhelm Eblinger to found a company named Benz & Cie, which produced industrial machines and grew to twenty-five employees by 1885. On January 29, 1886, the company filed a patent DRP-374325 on the Motorwagen or an "automobile fueled by gas." The first model was challenging to steer, which resulted in a collision with a wall during the first demonstration.

By 1888 Carl had improved the Motorwagen significantly enough that his wife Bertha was able to drive the vehicle from Mannheim to Pforzheim, a distance of about sixty-five miles. Bertha brought her two sons, Eugen and Richard, along for the ride to visit her mother, telegraphing Carl about their trip. Among the innovations Bertha helped her husband with was when she arranged with a shoemaker to nail leather onto the brake blocks, inventing the first brake lining after some terrifying downhill slides during the journey. What Bertha did not help her husband with was the tires for the Motorwagen, which were made of wood and the car likely had a rough ride. Benz's Model 3 made its debut at the 1889 World's Fair in Paris with about two dozen cars eventually being built.[1] Benz & Cie eventually merged with a company named Daimler Motoren Gesellschaft (DMG) in 1926 to become Daimler Benz. All of Daimler-Benz's cars were and are named Mercedes-Benz after the ten-year-old (in 1902) daughter of Emil Jellinek who had designed the 1902 Mercedes for DMG. Over the last century many innovations around tire production have contributed to vastly smoother rides than the original Motorwagen. The key components of a modern automobile tire include rubber, carbon black, and steel. Unfortunately, the production of carbon black is enormously bad for Mother Earth as it is currently produced by burning petroleum, which creates many noxious gases—nitrogen oxide, carbon monoxide, sulfur dioxide, and the one the world currently cares most about, carbon dioxide.[2] Monolith Corp. has developed a cleaner process for producing this key commodity that will save over 1 million tons of carbon emissions per year—equivalent to the annual emissions of over two hundred thousand cars.

This chapter presents efforts by individuals and companies to provide tires that are far less damaging to the environment. This description will move from upstream at the beginning of the automotive supply network, starting in Haslam, Nebraska, drive through the tire industry with producers such as Goodyear and Michelin, and end downstream with the efforts of Bolder Industries to create a profitable circular supply network for used tires.

Before starting the journey along the supply network, it is important to consider how universal automobile tires are in modern society and discuss the benefits they bring to humanity along with the many negatives. I won't spend much time on the benefits—most people love to drive—at least when not stuck in horrible traffic.

Did you know that *Billboard* has a list of the "100 Greatest Car Songs of All Time?" The last song on the list is "Drive" by the band Cars, which was released in 1984. The Eagles' "Take It Easy" from 1972 is at number 12 and Nelly's "Ride wit Me" from 2000 is at number 8. And at number one, we have "Born to Run" by Bruce Springsteen from 1975. However, despite the convenience and freedom that cars provide, they have a negative impact on the environment due to both the tailpipe emissions and particulate matter from tires. Worldwide over two billion tires are produced and disposed of each year. Waste tires present numerous hazards: rainwater accumulates in tire piles creating breeding grounds for mosquitoes, while piles of discarded tires catch on fire releasing toxic fumes visible from space. Kuwait has the largest tire pile in the world with 40 million tires buried in the desert. In October 2020 1 million tires caught on fire, releasing an estimated 10 million gallons of toxic chemicals before being extinguished days later. The tire is where the climate rubber meets the road.

Much of the focus for the auto industry in transitioning to EVs is on reducing carbon emissions, yet at the same time, tires emit many other toxic ingredients. Tests by Emissions Analytics, an English engineering consulting company found that emissions from tires consisted of more than four hundred compounds of different sizes and levels of toxicity. These emissions can be carcinogenic and are linked to heart and lung disease. Further, regulators have not tackled the issue of tire emissions, in sharp contrast with tailpipe emissions from traditional ICE vehicles. In Anaheim, California, researchers have found that brakes and tires account for 30 percent of harmful emissions, while exhaust emissions linked to gasoline were only 19 percent.[3] The chemistry involved in making tires that will perform well and last on roads and highways worldwide is very complex. Many of the problems are associated with the fossil fuel foundation on which tire production rests. Other problems are associated with the fact that 90 percent of the rubber comes from trees that only grow in tropical locations, primarily in southeast Asia. This makes for a near monopoly and a high risk of disruption.

One effort to develop alternatives is spearheaded by Goodyear Tire and Rubber Co., which announced a multiple-year program with support from the US Department of Defense in 2022. The program seeks to develop and certify tires produced with rubber made from dandelions which any lawn owner knows grow much more quickly than the seven years it takes a rubber tree to grow, thus dandelions provide huge potential environmental improvements for tires.[4]

Upstream Supply Network: Hallam, Nebraska, 2016—Monolith

The town of Hallam was platted in 1892 when the Chicago Rock Island and Pacific Railroad reached the site, with the first landowner, Jacob Schadd, suggesting the town be named Hallau after his birthplace in northern Switzerland. A transcription error replaced the u with an m. Today Haslam is a town of less than three hundred people about twenty miles south of Lincoln, the home of the Nebraska Cornhuskers. In

1962 the second nuclear power plant in the US opened and the town looked poised for growth. For some reason buried in the history books, the plant only used nuclear power for a single year before converting to coal.[5] In 2016 another new technological marvel came to the town in the form of Monolith Corporation.

The global carbon black market is estimated to grow to over $18 billion in 2029. This commodity material is an essential element in manufacturing automobile tires, making up 30 percent of the overall weight. In addition to tires, carbon black is used in a vast variety of rubber goods, including hoses, belts, seals, and gaskets, as well as in specialty coatings such as paint and mascara. In many ways it is the secret ingredient that is everywhere in our world.

The first use of carbon black was in 1910 when the BF Goodrich Company added it to rubber as a filler to extend the life of tires. Prior to this addition tires were not black; in fact, some early Tin Lizzies had white tires[6]—which is ironic because Henry Ford's most famous saying about the Model T is that you can "have it (the body) in any color . . . so long as it is black." Chemical engineer Jack Koenig, author of the book *Spectroscopy of Polymers*, explains the value of this dirty, magic substance: "Carbon black imparts strength and toughness to a tire as well as improves the rubber's resistance to tearing, abrasion, flex fatigue and also increases traction and durability. A tire would last less than five thousand miles without carbon black." Most new cars come with a four- to five-year warranty on the tires, meaning the tires are expected to last roughly ten times longer than those before carbon black.

This is where Robert Hanson, Pete Johnson, and William Brady enter the scene with Monolith, originally named Boxer Industries. Pete and Robert (Rob) met at Stanford while earning their master's degrees in mechanical engineering and then worked for Ausra Solar from graduation until launching Monolith, nee Boxer, in 2013. In June 2023 Rob granted me an interview and Austin Burk, plant manager, provided a plant tour of Olive Creek 1 the facility in Hallam.

> Pete and I are really the first two founders. We were both working at Ausra in 2007 to 2010. Then the company gets sold to Areva, which has

> maybe one hundred thousand employees and is 90 percent owned by the French government. Pete and I start talking and agree that nuclear is a great business, but we are thirty years old, we don't want to spend our career doing this.[7]

Power companies are generally considered some of the slowest-moving and least innovative companies in the world. Everyone needs the power they provide, yet the enormity of the capital investment required combined with challenges in shepherding new construction through the permit and approval process (the reader may recall the lengthy delays in getting new wind power plants built in the United States as described in chapter 8) make the industry appear stagnant at times.

> So Pete and I decide to start something, but we didn't start the way most people do, which is you know, they're a postdoc, and they're working on something in the lab. They're like, "I'm going to take this and turn it into a company." It wasn't like that. We started with two founding principles that our project needs to be *clean,* and it needs to be *cheaper*. Because if it's not those two things, it won't have the impact that we want to be economically viable.

The pair of founders visited national labs, universities and corporate R&D facilities. According to Hanson, "It's easy to find technologies that are clean and more expensive—they are a dime a dozen."

Hanson relates that the pair came across the work of Professor Laurent Fulcheri at the Ecole des Mines de Paris, which ironically is not situated in Paris but just north of Cannes. Johnson and Hanson flew to France to visit Professor Laurent in his lab where a kaleidoscope of rewired ideas bloomed. Mark Twain wrote in his autobiography: "There is no such thing as a new idea. It is impossible. We simply take a lot of old ideas and put them into a sort of mental kaleidoscope. We give them a turn and they make new and curious combinations."

At this point circa 2011, Fulcheri has been researching "new" methods of producing carbon black and hydrogen for well over two decades. Fulcheri and his coauthor Yvan Schwob published an article titled "From Methane to Hydrogen, Carbon Black and Water" in 1995. This article was published in the *International Journal of Hydrogen Energy*, which had been publishing research on hydrogen energy for two decades at the time. In my opinion, mostly informed by social sciences, this suggests that there is a role for hydrogen in decarbonizing planet Earth. The thing is—taking an idea from bench science in a laboratory to mass production and mass adoption is extraordinarily challenging and rarely, if ever, achieved by the efforts of a single person or small group of people.

As coauthors Fulcheri and Schwob wrote in the paper's abstract:

> From physical considerations related to existing processes, the authors present a *theoretical* study which could open the way to new plasma-assisted processes. More anecdotally, a certain number of natural gas resources remain unexploited due to their isolation; it is possible to transform these resources into carbon and water without any external energy supply. It would then be possible to irrigate the desert while producing a solid-state product whose transport may be easier than gas.

Here in the form of two idealistic and brilliant engineers educated at Stanford in the Silicon Valley, was the opportunity to move from theory to practice using the research of a two-decade career and the learnings built on the efforts of hundreds of scientists. However, to bring this technology to market required an extreme willingness to commit. According to Hanson:

> What Laurent had discovered was how to turn the knobs on this little reactor (that he had invented) just right, he is a great experimentalist. The carbon black that came out was the valuable stuff that we could sell for two dollars a kilogram, not coke which is only worth ten cents.

The two entrepreneurs now believed they had a concept that fulfilled their two fundamental principles of clean and cheaper. The only problem was that the venture capitalists were not buying what the two young dreamers were selling. Hanson continues:

> We want to do this only in America, with private equity and venture capital building around technology up to big scale. So we go [and] start

> pitching investors. And they're like, you know, "Cool idea. But you guys know nothing about carbon black; this is complicated. The sales cycle is multiple years." And so one of them eventually said, "We'd love to make an introduction to this guy who knows carbon black as well as anyone in the world, Bill Brady, who is the former president of Cabot, the biggest carbon black company in the world." So we got a zoom call set up, or I think it's actually a conference call, just a voice call. This was before COVID-19, so we're gonna have a voice-only call. And we're like, "Hey, this thing is going to make or break this business. Because, like if Bill gives the thumbs down, like that'll spread and we're done."

Passionate to the core, the pair bought tickets and jumped on a plane to Boston. For the first hour, Brady was pretty stoic, according to Hanson:

> You could tell, he's thinking *lots of piss and vinegar in these guys*—but they don't know how *hard* it is to sell this material that makes up a third of every tire and emits copious quantities of greenhouse gases. It takes something huge to make companies change their production methods.

In the end, the trio met for three hours, and the pair won Brady over with economic arguments, aka the Ability to Profit. The conversation with Bill was one of those "I can't believe I didn't think of it" moments. This is one of the most powerful forces in the world. There is a saying—*Do you have thirty years of experience or the same year of experience thirty times*? Pete and Rob were young and idealistic, indeed. At the same time, history finds that young people not beaten up by years of experience have the energy and idealism to pursue world-changing ideas.

The first hour of the meeting was fairly subdued, Bill clearly wasn't buying what Pete and Rob were selling. As Rob tells the story,

> That was the first hour of that meeting is Silicon Valley. Trained entrepreneurs coming in and being like, "We're going to disrupt the carbon black industry and have 100 percent market share in five years and built in and be like these guys don't know what they're doing." So after about an hour, he's kind of zoned you out. . . . Bill is clearly thinking, "These guys are delusional on the time frame!"

They were delusional on the time frame, but they also were right that the DNA of their idea was solid. Pete and Rob kept talking. They drew a basic input/output diagram. Inside the process box they listed the five biggest carbon black firms in the world, which are a near monopoly controlling roughly 80 percent of the world market. And Brady kept returning to: "It's the unit economics, baby. Take Cabot Corp. for example, its earnings for 2024 are projected to be ~$800 million with a stock value of 6.5 EBITDA. This values the entire company at roughly $5.2 billion or about $1.15 million per employee. Cabot is a commodity-focused company that has to fight like a dog for scraps of food. Compare that to Tesla valued at roughly $800 billion or about $6.5 million per employee. Much better and highly correlated with the passion customers feel for their products. I am confident there are people in the world who are passionate about carbon black, including Pete, Rob, and Bill Brady, yet many are more passionate about cars.

Returning to the discussion of unit economics, Pete and Rob drew an arrow into the process diagram representing petroleum input. At the time, the cost of oil to produce one ton of carbon black was about $1,000. Cabot and other carbon black companies made a profit, but not much, thus, the pair also drew an arrow at the bottom left, representing all other expenses—labor, maintenance, and so on—and labeled it as $500. After all this, Cabot was left with a slim profit. This is why in a capital-intensive, low-profit margin industry there has not been a new carbon black plant built since World War II. See figure 9.1 for a comparison of legacy and Monolith Carbon Black production processes.

Pete and Rob explained that using natural gas was far less expensive than using petroleum. This got Brady's attention: "So can't I just run natural gas as the feedstock to our plants at Cabot." No. Completely different chemistry and technology. It would require a lot of electricity, which is what Prof. Fulcheri had been working on and was Pete and Rob's fundamental point. It was also potentially disruptive because the United States had recently had a shale revolution and natural gas had become abundant and inexpensive.

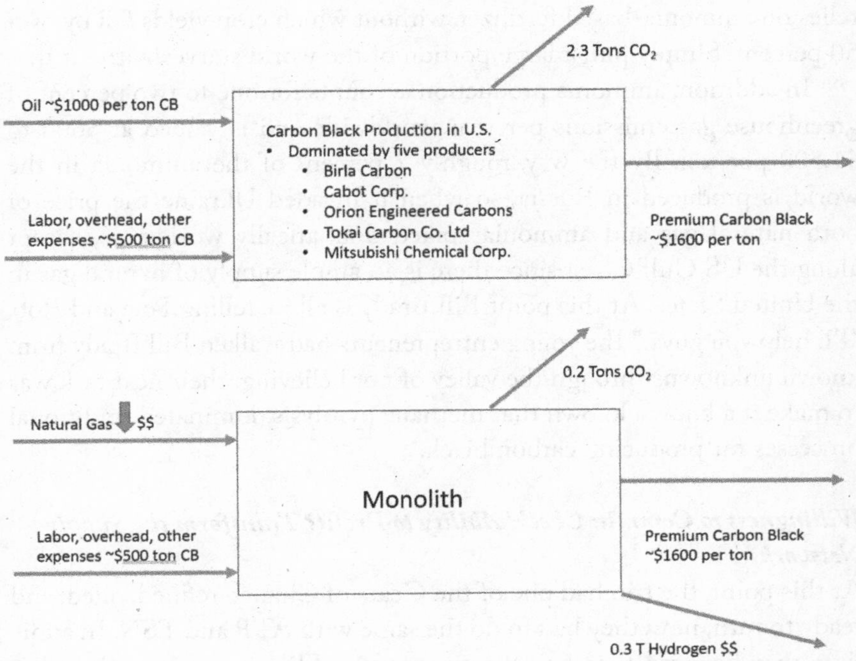

Figure 9.1. Comparison of Legacy and Monolith Carbon Black Production Process

The trio continued their conversation for two hours beyond Bill's allotted initial hour. The eureka moment was when the young entrepreneurs said:

> Hey, Bill, there's also this other part where we don't just make carbon. We also get hydrogen, and this is, you know, this is 2012. So like no one's talking about hydrogen for buses and trucks, it was just like in hydrogen's [or ammonia's] worth, you know, another thousand bucks, and that's just going into fertilizer.

Paydirt! Now they had captured Brady's complete attention. This indeed was potentially cleaner and less expensive. Ammonia production worldwide is approximately 235 million tons per year and is a component of almost all fertilizers. Approximately half the world's food supply relies on ammonia-based fertilizer, without which crop yields fall by over 50 percent. Simply put, a large portion of the world starves without it.

In addition, ammonia production accounts for one to two percent of greenhouse gas emissions per year worldwide and is valued at $600 to $1,500 per ton. By the way, roughly 7 percent of the ammonia in the world is produced in Russia, so when it invaded Ukraine the price of both natural gas and ammonia spiked dramatically worldwide, except along the US Gulf Coast since there is an ample supply of natural gas in the United States. At this point Bill Brady is all in, telling Pete and Rob, "I'll help you guys." The young entrepreneurs had walked Bill Brady from known unknowns through the valley of not believing; their next task was to make it a known known that methane pyrolysis dominated traditional processes for producing carbon black.

Willingness to Commit: Check! Ability to Profit? Transform the Supply Network?

At this point, the trio had one of the Gears of Change refined, oiled, and ready to turn; now, they had to do the same with ATP and TSN. In addition, they believed they had the answer to ATP, the rest is engineering and working with supply partners or the TSN gear. So off to work they went, in Rob's words:

> The economics of the Methane Pyrolysis process become the center point. Thus we were there [willing to commit] and those other two things are absolutely critical [ability to profit and transform the supply network]. But it's like what we didn't want to do is have something where we work so hard over our whole careers, right? Because these will take whole careers bringing them to full maturity. And then in the end even though you're able to convince everyone and get everything done you're more expensive.

Since this book focuses on supply networks, I will omit most of the efforts that all Monolith supporters engaged in from 2012 to 2016. It is an understatement to say that there were many moving parts to finding funding to build the firm. But clearly, in hindsight, these were found. In terms of TSN, the next step was proving that Prof. Fulcheri's small reactor would work on a larger scale. As Neil Armstrong famously said, "One small step for Man, a Giant one for mankind." And giant steps are *tough*.

The first public news on Monolith appeared May 18, 2015, in *Chemical and Engineering News* with Michael McCoy writing, "Redwood City–based Monolith Materials wants to build the first new carbon black facility in the US in some 30 years."[8] In breaking this story, McCoy opined that it was an intriguing idea that had been tried before and met with failure. In the 1990s, a Norwegian engineering firm Kvaerner (later renamed Aker), innovated a process to turn methane into carbon black and hydrogen using plasma pyrolysis (another name for Methane Pyrolysis). Kvaerner built a plant in Montreal, yet shuttered it by 2003. While it has not been publicly announced, a paper presented at the Electrochemical Society meeting in 2008 postulated that the plant encountered problems with "electrode wear and carbon bridging between electrodes." In plain English, the huge torch needed to catalyze the methane—had breakdown issues.

This is one of the small but subtle differences between engineering and bench science. If something has been done at a small scale, it can usually be done on a much higher scale—that is engineering. Bench science is developing a theory and showing it is possible—as Prof. Fulcheri did between 1995 and 2010 when Pete and Rob found him. Modern energy (and tire) engineering and production is a capital-intensive game where radical/disruptive innovations require tons of capital—and patience. Bill Brady said of investors Kern Partners and First Green Partners, "These are big, serious private equity investors (with the ability to be very patient and persistent)."[9]

Furthermore, in case the reader is thinking that we only care about carbon emissions, there are many other environmental challenges associated with carbon black (and hydrogen) production. Carbon black production also emits particulate matter, with Cabot Corp. agreeing to spend $84 million to mitigate particulate matter emissions in Louisiana and Texas in 2013. While Monolith claim(ed) that their new process will also reduce sulfur and nitrogen emissions, there was substantial doubt. Rob Hanson countered concerns at the time by pointing out that the pilot plant was located only fifty yards from San Francisco Bay and that environmental authorities approved the facility. To shorten the story, the Seaport facility in Redwood City operated without major incidents until 2015, at which point Monolith announced the next phase as in investment in its first commercial facility, named Olive Creek 1 (OC1) in Hallam, Nebraska.

Moving the clock hands forward quickly, Monolith successfully built the OC1 facility and in 2021 and 2022 began limited delivery of product to customers, including Goodyear, the biggest tire company in the world and an icon of American branding. Two key inflection points for the corporate pair came first on December 9, 2021, when the tire giant announced it had signed a letter of intent with Monolith for development and potential use of carbon black from methane pyrolysis. The second inflection point came in May 2023, ironically exactly eight years following the announcement of Monolith's move to Nebraska. Goodyear's Chris Helsel (senior VP global operations and chief technology officer) argued that "the use of carbon black produced by methane pyrolysis is an example of how we are collaborating with our suppliers, like Monolith, to utilize sustainable materials in our consumer products without compromising on performance and safety."[10]

Now, in addition to Rob Hanson, Pete Johnson, Bill Brady, and everyone associated with Monolith, Goodyear, a perennial Fortune 500 member, was indicating high WTC. Companies in the Fortune 500 can't afford to sacrifice ATP, so I turn now to an examination of key supply chain decisions that lubricate the remaining gear.

Location, Location, Location

On first glance the heading may seem like a realtor's classic mantra about selling a house quickly and profitably, but location is also a critical structural supply chain decision. This is particularly true in a capital-intensive industry in which a company like Monolith is seeking to disrupt decades of stasis. When corporate leaders began to look for a location to build a production-scale facility, they knew it couldn't be in California. Methane pyrolysis needed a lot of electricity. California's average cost of electricity is among the highest in the continental United States at 19.65 cents per kWh. This is more than 2.4 times Idaho, the least-costly state at a little over 8 cents per kWh. Easy decision? Not quite.

Monolith also needed a way to utilize low-carbon energy, requiring a state with plentiful renewable energy. Alternatively, they wanted a location where Monolith might connect with the utility provider to feed its own clean, low-emission carbon as input to the energy production process. In this case, a location close to a willing power partner.

Two other considerations were paramount. First, the chosen site also needed ready access to a plentiful source of natural gas. Second, they needed access to rail lines for shipping the finished product—ammonia, hydrogen, and carbon black. This is a challenging list to achieve. The Monolith team collected data and carefully analyzed almost one hundred sites.

Many things about California are attractive, particularly the weather. The state also has a mix of ocean and mountains. Nebraska has neither, yet turned out to be perfect for Monolith. As the photo in Figure 9.2

Figure 9.2. Photograph of Monolith Facilities in Hallam, NE

illustrates, the chosen location has all the key elements. While all are fundamentally important, first among equals was the power source as announced on April 17 and reported by Sonal Patel. Readers may remember earlier in this chapter Hallam was described as possessing the second nuclear power plant in the United States, one that was inexplicably converted to coal within a year. The Monolith announcement came with an explanation:

> In what could be an industry first, Nebraska's largest electric utility plans to replace an existing coal-fired boiler with one that uses hydrogen fuel. The Nebraska Public Power District (NPPD) said on April 17 that it plans to fuel its Sheldon Station plant in Hallam, Neb., with hydrogen produced by Monolith Materials as a co-product of its production of carbon black using natural gas as a feedstock. NPPD anticipates that the boiler conversion will equip the Sheldon Station with 125 MW while slashing air emissions. NPPD President and CEO Pat Pope noted that the Sheldon Station has always been (a leader) in innovation: It was built as an experimental nuclear plant in 1963 for the Atomic Energy Commission, but later decommissioned. Part of the plant was later converted to combust coal. Today, its two boilers have a capacity of about 225 MW. The planned boiler conversion at Unit 2 could make it the "*first utility-scale hydrogen powered generator*," said Pope." The project will require that Monolith build a new manufacturing facility near the Sheldon Station. The companies expect to break-ground on their respective operations in 2016, with an expected completion date of 2019, NPPD said.[11]

COURTESY OF ROB HANSON

Here we see Monolith transforming the supply network in partnership with NPPD. Refitting the powerplant to produce electricity using hydrogen is far from a simple or inexpensive process, at least in terms of capital investment. Leading up to this announcement, there was undoubtedly a negotiation process where Monolith likely agreed to underwrite some of the capital costs of transitioning to hydrogen. This is how organizations work in supply networks to achieve a win-win rather than facing off in an adversarial negotiation where one side comes out the "loser."

Also shown in the photo is that the electrical substation is very close, which makes the electricity flow right for methane pyrolysis. Similarly, Monolith could easily tie into a rail line and a natural gas pipeline. The second key supply network effort involves the process of producing clean hydrogen and carbon black.

Process, Process, Process

In a useful simplification, the literature on operations (supply network) strategy holds that a business can seek to gain a competitive advantage in one of two areas: innovating their products or in their processes. It is extremely rare for organizations to be better than average let alone top 10 percent in both. Such companies as 3M, IBM, and Toyota are like switch hitters in baseball, a threat from either side of the plate, thus premiere athletes/corporations.[12] Toyota became the largest car manufacturer in the world on the back of operational excellence which is often also called the Toyota Production System. Within the automotive industry, Toyota's production *processes* have been and remain world-class. A reasonable argument can be made that for the first ten years of the twenty-first century, Toyota was also leading in *product* innovation. In particular, Toyota first sold the Prius in Japan in 1997 and worldwide in 2000. Reviewed by Edmunds in 2007 as "too small, too slow and too conservatively styled to get much attention outside the hard-core environmentalist community," despite such negative reviews, Toyota had sold over a million units by May 2008 when Muskla was just debuting the Roadster. If Toyota had been more agile in investing in developing a more stylish model and marketing to a broader audience, the door that Musk and Tesla walked through might never have opened. Rather than leading in product design, Toyota's strikeout is what led to the current EV race.

Returning to Monolith, in 2012 Rob, Pete, and Bill were confident that they had a substantial product advantage—cleaner hydrogen, ammonia, and carbon black—and *potentially a superior process*. The challenge would be to develop the process.

Most radically innovative processes are built in stages that expand strategically over time. Companies typically fail when they try to achieve a "moon shot" and make a huge process jump without staged process innovation such the Mercury and Gemini programs that preceded Apollo. Consider Webvan. Founded in 1996 during the dot-com bubble, the company was financed by venture capital companies Benchmark Capital, Sequoia Capital, and Louis Borders who cofounded Borders Books in the 1970s. Webvan raised an aggregate total of over $750 million by the end of 1999, valuing the company at almost $5 billion. The company planned to deliver customized orders of groceries to customers within a thirty-minute window and hoped to expand to twenty-six cities by 2001. Pressured by investors, the company placed a $1 billion order with Bechtel to build its warehouses and bought a fleet of delivery trucks. Expanding *before* they had developed a reliable and predictable process of fulfilling thousands of orders was one of several monumental mistakes. Ultimately, Webvan lost over $800 million while selling less than $300 million in groceries. CNET named it one of the largest dot.com flops in history.[13]

The plan was to build the Olive Creek facility in two stages. The first stage dubbed Olive Creek 1 or OC1, is a proof of concept that carbon black can be produced with methane pyrolysis to the quality and cost specifications that downstream manufacturers require. Theoretically, this could be done, but in practice there were thousands of engineering challenges. When I had the chance to talk with cofounder Rob Hanson, we quickly dove into a very animated conversation as my first job out of college was working on nuclear submarines in 1990. Much of the technology that Monolith is employing is based on the mechanical equipment for power generation that has been in existence since Robert Fulton first took passengers up the Hudson River on the *Clermont* in 1807. At the same time, there were many things to be invented.

Perhaps the biggest challenge was the plasma torch that heats the methane. The reactor that Professor Fulcheri showed Rob and Pete when they first met him ran on between 100 and 300 hundred kW of electricity. OC1 required a 16 MW torch, when the largest commercially available torch at the time was *1 MW*. Putting this in perspective, the existing 1 MW plasma torch used enough power in one hour to drive my Tesla 3,533 hours. In other words, a lot of electricity. When Monolith set out to invent a 16 MW torch, the task was extraordinarily complex. At a minimum, this would be enough to power almost a thousand Tesla's to drive sixty miles for every hour it operated. Further, this was unproven technology, so while the Monolith team believed it could be done, they also knew it was going to be extraordinarily difficult.

The Monolith website has a beautiful illustration of the methane pyrolysis process. Looking at the illustration makes it look magical, yet the reality is this is an enormously complicated process to manage. The second fundamental challenge to overcome is developing a plasma torch that is reliable and can be counted on to have far more uptime than downtime. An article published in 2020 in *Chemical & Biological Engineering Reviews* states:

> In plasma decomposition, high local energy densities and temperatures of up to 2000°C are generated by means of a plasma torch. Large gas volume flows are usually recirculated to stabilize the plasma. In the area of the actual plasma torch, cooling, electrode wear, and carbon deposits are among the greatest technical challenges.[14]

In plain English, keeping the torch running is really hard. As I write this, Monolith has applied for or holds over one thousand patents. Corporate leaders have proven they can produce at scale, and they are planning to begin construction of OC2.

Two other elements of the Monolith methane pyrolysis process potentially provide a competitive advantage. First, the process designed for Olive Creek 2 has high flexibility. OC2 can handle two major inputs, either natural gas of biofuels such as ethanol or fuels made from waste products from agricultural or forestry production. These include rice straw, risk husk, wood chips, and saw dust. The photograph in figure 9.3. shows Rob Hanson explaining the chemistry of this process to me in June 2023. Theoretically, when biofuels make up the input material, the entire process is carbon-negative—meaning it is *capturing carbon from the atmosphere*.

Another noteworthy design feature is the use of two separate lines of reactors for methane pyrolysis. The intent here is to be able to run five

Figure 9.3. Rob Hanson Draws the Chemical Reaction for a Negative Carbon Emission Process to Produce Carbon Black

of the six reactors in each of the lines in parallel for high-volume production. The sixth reactor for each line is essentially a spare/backup. This addresses the challenge identified in the article in *Chemical & Biological Engineering Review* earlier. In essence the plasma torch and reactor are the key bottleneck in the process because of the technical complexity inherent in this element.

On the output side of the planned process for OC2 are the products Monolith will sell. Carbon black will be a mainstay, with a fairly complex back-end process as shown in Figure 9.4. As noted earlier, refining carbon black that meets the exacting standards of a Goodyear or competition tire manufacturer involves some precise chemistry, thus, OC2 is designed with a flexible back-end process where the equipment can be adjusted for different outputs. In addition, to various grades and compositions of carbon black, the process can be set to produce either hydrogen for power generation elsewhere (starting with feeds to NPPD in a virtuous cycle) or ammonia for fertilizer. This flexibility allows Monolith the opportunity

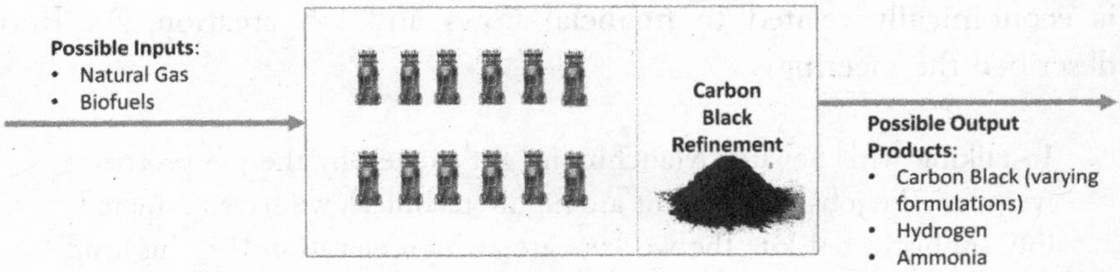

Figure 9.4. Flexibility in the Methane Pyrolysis Process

to adjust product mix according to market prices and costs for its inputs and outputs, which is a tremendous advantage. In sum, I like Monolith's chances of achieving economic and climatic success.

In addition to the ability to align the Gears of Change for success, the company has also found strong political supporters. Senator Joe Manchin (WV) has made a name as a critical vote in the US Senate given the balance of representation in this august chamber. Leah C. Stokes, an associate professor of environmental politics at the University of California Santa Barbara is a Manchin critic, writing on July 16, 2022:

> Over the past year and a half, I've dissected every remark I could find in the press from Senator Joe Manchin on climate change. With the fate of our planet hanging in the balance, his every utterance is of global significance. But his statements have been like a weather vane, blowing in every direction. *It's now clear that Mr. Manchin has wasted what little time this Congress had left to make real progress on the climate crisis.*[15]

In publishing this scathing opinion, Dr. Stokes referred to the stripping of over $500 billion in climate investments from the Build Back Better bill passed earlier in 2022. She had a point. At the same time, many have said that politics is a blood sport in which the participants are battling for competing objectives. As has been covered earlier in this book, when the Inflation Reduction Act passed the Senate in August 2022, it was with Senator Manchin's vote, and it did so with many hundreds of billions of dollars focused on cleaner energy of the type Monolith is seeking to provide.

As Rob Hanson described in a recent meeting with Senator Manchin, a significant element in his support of the IRA and Monolith is economically related to financial flows and job creation. As Rob described the meeting:

> In talking with Senator Manchin, he got excited by the job prospects we offer. The jobs in the plant are highly technical, we have an incredibly sophisticated kit. The workers are at high elevation [i.e., high off the ground] to perform maintenance and changeover—all types of risks, even while Monolith takes all proper safety precautions. Senator Manchin basically said, "These are the types of jobs coal miners want. Mechanically inclined, hardworking, get your hands dirty. Not coding—this is an appealing, dignified, well-paying job."

In sum, I wasn't there, but it seems Senator Manchin is supportive of Monolith's mission because of the ability to bring well-paying, dignified jobs to Nebraska, a very red state that voted 58.5 percent Republican in the most recent presidential election.

Midstream in the Process: Goodyear Joins with Monolith

The previous examination of Monolith and carbon black illustrates a bold investment to deliver a product that makes up 25 to 30 percent of every automobile tire. Further, roughly three-quarters of the carbon black produced worldwide is used in tires. From here, I turn to the next step in the production process, tire manufacturing, focusing on Goodyear, one of the largest tire companies in the world. Headquartered in Akron, Ohio, the "Rubber Capital of the World," Goodyear generated 2020 revenues of over $12 billion through the efforts of over seventy thousand employees with fifty-seven tire production facilities and over twelve hundred tire and auto service centers.

In the 1920s, Akron was a center of innovation due to the many chemists, engineers, and tire experts who developed the tires that Ford, GM, and Chrysler used to create the American automobile market and sell millions of cars. To provide an idea of the scale of possibilities for supply network transformation, Akron nearly tripled in population from 1910 to 1920 as the tire industry and Goodyear grew in lockstep with Ford and Detroit. Columbus, the state capital of Ohio, was roughly the same size as Akron a century ago. Over the last five decades, sadly, Detroit and Akron have been devastated economically, with Detroit losing 60 percent of its population and Akron losing almost a third. Columbus, by contrast, has nearly doubled in the same time span, mainly because it has been blessed with a high mix of services, particularly financial-service-based companies. At the same time, the marriage of greater sustainability with continued profitability in tire production today is supported by a vast community of innovative professionals with diverse skill sets.

Tire manufacturing is an extraordinarily challenging business. On the one hand it deals with a product that combines numerous materials that have a colossal diversity of chemical properties with a manufacturing process that must meld these materials together to form a tire that operates safely and with good performance in a wide variety of driving situations. At the same time, tire manufacturing is a highly capital-intensive business with generally low per-unit profit margins. Thus, radical changes to the inputs or processes are rare and challenging to pull off, as the early profile of the discussion between Rob, Pete, and Bill Brady illustrates.

In December 2021, a major announcement from Goodyear noted that the tire giant had signed a collaboration agreement and letter of intent with Monolith. Chris Helsel, senior VP of global operations and chief technology officer, has noted: "At Goodyear, we're committed to sustainability and making a positive impact by our choice of the materials we use. Monolith is one example of how we use sustainable materials in quality products that deliver a better future."[16]

The development process for certifying a new material and then scaling production of that material to profitability is an extraordinary collaboration between organizations. Most likely the process of reaching the initial collaboration agreement required a couple years of ongoing talks and negotiations. From that point an additional eighteen months passed before a joint announcement in May 2023 that Goodyear would produce and sell an ElectricDrive GT tire, specifically tuned for EVs employing a new tread formulation and Monolith carbon black. The next few paragraphs draw on my several interviews with members of Monolith's sales team.

Monolith has hired several dozen people with decades of experience in the plastics, chemical, and carbon black industries to help build relationships with manufacturers, including Goodyear, Bridgestone and Michelin. In addition, there are hundreds of other companies involved in both the automotive supply network and other goods including paints, plastics, and apparel that employ carbon black as a key ingredient in their end products. The tire manufacturers have entire teams of experts in diverse scientific and technical areas to manage the certification process. The process of developing and certifying a new grade of carbon black for high volume tire production is complex:

> Carbon black is not just 30 percent of any automobile tire; I call it transportation—it's in large Caterpillar construction tires that have treads and rubber. It also provides ultraviolet stability. It provides jetness, which allows it to provide different shades of black, brown, yellow, blue.[17]

In the carbon black industry an analogy is commonly utilized comparing carbon black to grapes and the many thousands of products it is employed in, such as wines. Think of carbon black like a grape or a cluster of grapes. You have the individual particle/grape and then you have the cluster. Within grapes you have different varieties, small grapes with different clusters, then there are Welch's grapes with a different shape. Carbon black is similar; each batch of carbon black has ten to fifteen different chemical properties that need to be measured. Thus the carbon black producer (Monolith or Cabot) needs to run an in-house lab, while the tire manufacturer (Goodyear or Michelin) also runs an in-house lab. Furthermore, the chemical and physical properties of the input carbon black change when combined with rubber, steel, and silica to form a finished tire; thus there is a large amount of testing at the end of the production line.

Carbon black can be used in a vast variety of products, including automobile tires (high safety concerns), jet airplane tires (much higher safety concerns), roofing products such as shingles (long-term quality concerns as shingles are expected to last twenty to thirty years) and mascara (cosmetic safety concerns). A non-automotive company, Advanced Drainage Systems (ADS), employs carbon black to turn recycled plastic into a consistent shade of black. ADS is a leading manufacturer of sustainable stormwater and onsite septic wastewater solutions. With over $3 billion in revenue in 2023 and approximately sixteen thousand customers in North and South America, ADS is the largest user of recycled High-Density Polyethylene (HDPE) in the United States. HDPE is made from petroleum and is commonly used in containers for milk, motor oil, shampoos and conditioners, soap bottles, detergents, and bleaches. In fiscal 2023, ADS consumed 25 percent of all recycled pigmented HDPE bottles in the United States, which is equivalent to 650 million pounds of HDPE or taking sixty-three thousand cars off the road.[18] Thus, ADS operates one of the more successful examples of a circular economy in the world, which can provide insights for aspiring companies in the quest to master circularity for Li-ion batteries as examined in chapter 8 and companies like Bolder Industries, which will be presented at the end of this chapter. ADS pipe products are utilized in stormwater systems designed to last for over a century, thus, possessing strong protection against ultraviolet light is of critical importance. "While the pipe is waiting to be buried it may sit in the yard for three months, six months, up to a year. You can't have UV degradation. Carbon black provides UV stability that is critical."[19]

At tire manufacturers, there are teams conducting initial material specification qualification, plant teams at each production plant, and teams in the field. In many ways, representatives of carbon black companies including Monolith can be considered similar to the quarterback on a football team. There are two teams working together [Monolith and Goodyear] and an upstream carbon black expert is helping align all the different functions. As with a football team, if one team member runs the wrong play you have a busted play. Except in tires, just like the ignition switch situation described with GM earlier, mistakes can cause accidents and death. Unlike making wine from grapes, where a bad product won't sell or can be spit out, defective tires can be deadly.

There are hundreds of suppliers throughout the automotive industry and many other industries that utilize plastics and chemicals where carbon black is a critical ingredient. In fact, it is one of the top fifty chemicals used worldwide and a $12 billion industry. Most suppliers are seeking to get their "new and improved" products adopted downstream by a major company, whether it is a tire, automobile, or consumer packaged goods manufacturer. For these companies, it is a challenge to convince industrial customers to try a new ingredient as the risks appear more daunting than the rewards. This is similar to the grocery industry, where people with their side hustle try to convince Kroger, Publix and HEB to put a new product on the shelf. But grocers already have twenty to thirty thousand SKUs and can try a few new ones relatively quickly, and new food products aren't likely to kill anyone. These new tire ingredients usually have zero chance of making it past gatekeepers. Yet as was profiled earlier in the chapter, occasionally the right person sees an economic pathway toward making a more sustainable product, as Bill Brady did over a decade ago.

An off-the-record interview with a recently retired Goodyear executive provided some interesting insights presented as representative of the entire tire manufacturing industry. Alyssa Young (name changed) described how developing and certifying either new materials or new tires is a joint process between the material supplier and the manufacturer. Tires A and B may require different specifications. Overall Young's informal estimate was that the R&D and testing needed to get a new tire to the market represented about 10 percent of total costs. Elaborating on the partnership with Monolith, Young went on the say, "None of the businesses wanted to own the new carbon black and resulting tires," as the substantial upfront costs would be a heavy anchor on that strategic business unit's revenue and profit targets. Thus Goodyear initially looked to run the partnership with Monolith similar to its racing tire business, as an entirely new SBU with different key performance indicators. According to Young, "One of the businesses agreed to take the Monolith project on but also negotiated some wiggle room in the SBU's revenue and cost targets with the company chair. Our approach at Goodyear was to prove the concept in a single application, then scale. In contrast, others in the industry [Bridgestone? Monolith?] tried to scale faster by innovating with the Monolith carbon black across their broad product ranges." In either strategic approach, Young's primary point was that the tire manufacturers were committed to taking some risks in the short run with revenues and profit to bring a potentially industry and world changing tire to market.[20] In her words, "Is Goodyear going to compete on carbon black or is the entire industry going to adopt this?" Early signs are the industry as a whole is very willing to commit.

Endstream? Seeking a True Circular Tire Network

Bolder Industries was founded in 2011 as a technology-agnostic project developer focused on reducing landfilling and environmental impacts. It began operating its first tire recycling facility in Maryville, Missouri, in 2014. Ironically, this location is about two hours due east of the Olive Creek facility built by Monolith in Nebraska. Like Monolith, Bolder Industries also uses a form of pyrolysis to create its BolderBlack product. In Bolder's process, temperatures of approximately 1200°F in an oxygen-starved environment are employed to siphon off oil and gas from used tires. The dark solid left behind is then reduced to an ultra-fine powder that is then mixed with water to form tiny pellets for shipment to customers. Bolder is very similar to Monolith in that it employs a process that requires heavy investment to innovate and engineer. It acquired the Maryville facility in December 2014, building reactors in three stages. First came a pilot reactor to test and prove the concept, with the first commercial sales of its BolderBlack (carbon black) and BolderOil to Bruckman Rubber, which is now Cabot Plastics, an SBU of Cabot Corporation that Bill Brady Monolith cofounder previously was president of. The first production capacity reactor went live in February 2019 and the second in mid-2020. Together the two reactors required over $30 million to design, build, and bring online.

Much like Monolith, Bolder had to engineer solutions to many problems, with the pyrolysis process for converting used tires to usable commodities representing the DNA of the organization. Thus the receipt of a patent protecting critical feedstock in September 2022 represented the completion of a critical wall to ward off competitors. The company claims it recaptures 98 percent of the used tire to generate four products:

- BolderBlack which is a competitor product to the carbon black produced by Monolith.
- BolderOil, a sustainable petrochemical employed in solvents, renewable fuels, carbon black oil, and as a feedstock in circular chemical manufacturing.
- Gas, which is used for on-site power generation to reduce carbon emissions.
- Steel, which is sold to existing recycling firms.

The company projects that by 2027 it will have diverted over forty-five million tires from landfills, cut 1.8 million tons of CO_2 emissions equivalent to over three hundred thousand car years, reduced electricity usage by 400 million kWh and saved 3.6 billion gallons of water. Impressive? Certainly. Achievable? That is the key question.

As I write this, Bolder has signed contracts with key players across several industries. On the customer side the company had inked deals for Bolder Black with one of the ten largest tire manufacturers in the world and a Fortune 500 producer of paint with revenues approaching $20 billion annually. For BolderOil, the company had deals with one of the largest chemical producers in the world, with annual revenues north of $50 billion, and a large manufacturer of professional cleaning tools. Thus the downstream demand from customers appeared to be in a very healthy situation. Upstream, the company had agreed to a deal with Liberty Tires in July 2022 to gain access to feedstock—that is, tires to feed into its reactor.

Locking up a deal to acquire feedstock was critical to Bolder's success since operating reverse supply chains to close the circle is substantially more difficult than operating the forward supply chain. This is because shipping tires is expensive, so expensive that the manufacturing companies routinely ship tires from distribution centers in a semi-trailer packed like a giant rubber Tetris game. In other words, tires are packed into the semi in any way which they will fit—wall to wall, front to back and floor to ceiling. Approximately seven hundred tires per truck are manually loaded—using a human being which is very, very physical work. And this is the forward supply where tires are new and highly valuable.

Returning tires to a processing facility cannot involve long distances. The economics don't allow it as used tires are worth less than 20 percent of new; thus moving them very far is prohibitive. Founded in 2000, Liberty has grown over two decades, including over thirty facilities picking tires up from almost twenty thousand locations. With the largest network of recycling facilities in the world, Liberty has agreed to provide 3 million tires per year to Bolder's second facility in Terre Haute, Indiana, with an increase to 6 million when demand and capacity for Bolder's products merits it. So what is the difference in how these two firms handle the tires?

Liberty is upcycling, which is taking materials and transforming them for new uses. In this case, the primary use is turning the rubber into mulch, which can make playgrounds safer for children with a fall height rating of sixteen feet for products by IMC Outdoor, Liberty's newest facility in Hebron, Ohio. So why sell used tires to Bolder? Because Bolder's product is both higher value added and better for the environment. By recycling, which takes old materials and breaks them down into new materials that can be used for other products, Bolder creates higher-value end products that outperform Liberty's upcycled ones. Namely the reduction in CO_2 emissions for each tire fed into Bolder's facility is six to eight times that for Liberty. The value of the product that Bolder pushes out of its facilities is up to ten times what Liberty can produce.

One of the keys to reuse, whether recycling or upcycling, is capturing the used tires in the first place. Here the industry benefits from being a business-to-business operator versus a business-to-consumer, since its nearly twenty thousand tire-collection facilities all have the same fundamental challenge. Namely, what to do with used tires, which most commonly are removed from the cars they service in pairs or quads. A tire shop like Discount Tire either has to pay to have the tires hauled away and dumped in a scrap yard or receive a small fee from a company such as Liberty—easy decision, right? End consumers would be much better at either recycling or upcycling if there were a financial upside. The state of Michigan passed a law in 1976 requiring a ten-cent deposit on many glass and aluminum bottles. The recycling rate for the state fluctuated between 96 and 99 percent for two decades from 1990, beginning to trend down after 2010—likely because inflation ate away at the value of a dime. Prior to the COVID-19 pandemic the state's recycling rate was still close to 90 percent. In 2020 the rate fell below 75 percent and has not recovered. The good news for Liberty Tire? Tire replacement stores like Discount Tire accumulate used tires at a fairly predictable rate, with its over one thousand stores averaging between seventy and one hundred tires per day. Thus managing the route process for picking said tires up for transport to one of the over thirty processing centers is not overly complicated, provided the distance traveled is kept to a minimum.

The limitations regarding distance of transporting used tires don't disappear for Bolder. This is why Bolder built a single facility in Missouri, followed by a second in Indiana and a third announced for Antwerp, Belgium, in early 2022. Presuming the business can be run profitably, it is likely that Bolder will build other facilities strategically across the United States and Europe. But first, it must manage partnerships across the supply network. According to founder and CEO Tony Wibbeler:

> Cross-industry collaboration is critical to making the massive leaps needed to achieve our vision of transforming manufacturing sustainability worldwide. We're thrilled to formally partner with our friends at Liberty Tire who share in our vision and are right there with us, doing the hard work it takes to shift a massive supply chain toward long-term sustainability that will have positive ripple effects for decades to come.[21]

Bolder appears to be well down the path of transforming the supply network with Liberty Tire's help, yet there is at least one more significant piece to the puzzle. Bledsoe Innovation Group is Bolder's in-house laboratory to develop custom formulations and develop a database of approved compounds (i.e., oils and carbon blacks) that are approved for use and meet the standards set by end customers such as Pirelli or BASF. That lab is directed and managed by Archie George, who has a strong Ohio connection. Archie enrolled at Ohio State University in the 1970s to play football for Woody Hayes. The problem was that he wasn't talented enough for football, as Coach Hayes told him to protect a portion of the Buckeyes bench. A smart young man, George transferred to another university and earned a bachelor's in chemistry, which he utilized during a long career with Dunlop and Bridgestone before being recruited to Bolder. In 2019, George and his team determined that the feedstock the company was receiving had a higher proportion of silica, which was unacceptable to potential customers. The source of the silica was traced to a specific brand of tire which in turn prompted Bolder to develop *Patent No. 16/658,049—System and Method for Pelletizing Carbon Black Reclaimed from Waste Tires*, which was filed with the US Patent office in October 2019. The patent protects the production of BolderBlack through the use of a specific feedstock mixture comprised of passenger car tires, semi-trailer truck tires, and agricultural/off-road tires to produce a recovered carbon black with silica content below 5 percent and sulfur content below 20 percent. Thus Archie George and his team became starters for Bolder's team after warming the bench for Woody Hayes in Columbus.

Having discovered the work being done by Monolith, tire manufacturers such as Goodyear, and recycling partners Bolder Industries and Liberty Tire, the reader may think that this represents excellent progress toward solving the tire waste problem. John Sheerin, director of End-of-Life Tire Programs at the US Tire Manufacturers Association (USTMA) argues that the industry is in a state of transition and that the fundamental challenge is that the growth in scrap tire generation has exceeded the growth in scrap tire recycling markets. For context, in the USTMA's 2021 Scrap Tire Management Summary, the data show that scrap tire recycling is not keeping pace with increased annual generation. In total, markets consumed 71 percent of scrap tires in 2021, a decrease from around 76 percent in 2019. Thus, Sheerin argues for further action by US lawmakers to motivate more individuals like Rob Hanson and Archie George, and companies such as Monolith, Goodyear, Liberty Tire, and Bolder Industries to be willing to commit to true recycling and upcycling.

Chapter 10

Driving It Home

As I wrap the book, I am reminded of the unexpected friendship between Henry Ford, a pioneer in the automotive industry, and the renowned naturalist John Burroughs, as well as Thomas Alva Edison. In 1912, Burroughs published an article criticizing Ford's work, claiming that the popularity of gasoline-powered vehicles would harm Americans' appreciation for nature's beauty. Upon reading the article, Ford contacted Burroughs, and their mutual admiration for Ralph Waldo Emerson's philosophy of self-reliance led them to embark on a road trip in a Model T to visit Emerson's home in Massachusetts in 1913. This trip solidified a lasting bond between Ford, the fifty-year-old industry giant, and Burroughs, the seventy-six-year-old naturalist. Later, in 1915, Edison was also included in the group when the Panama-Pacific International Exposition dedicated a day to honor him.

Ford was able to convince Edison to join him and Burroughs on a road trip from one end of the country to California thanks to their existing friendship. While the inventors opted for train travel, their children Edsel Ford and Theodore Edison made the journey separately by car. Similar to Bertha Benz's historic first gasoline-powered road trip in Germany three decades prior, the younger generation had to overcome challenges such as poorly maintained roads, limited access to fuel and food, and the need to camp out at night. When the young men discussed their epic journey with their fathers in San Francisco, the inventors caught the bug, planning a trip in 1916 that would add Harvey Firestone to the traveling party. In 1906, Ford had selected Firestone's company to be the sole supplier of tires for its automobiles. Together Firestone and Goodyear were the largest manufacturers of tires in the United States for over seventy-five years until Firestone was acquired by a Japanese company, Bridgestone, in 1988. The trio of Edison, Firestone, and Ford were widely considered to be the titans of American industry, similar to today's tech brothers: Musk, Page, and Bezos. But the "Vagabonds," as Edison, Firestone, and Ford named themselves, did not have private jets to travel in, so they utilized Ford's most famous product to travel and experience nature, as profiled in Wes Davis's delightful book *American Journey*.

The group traveled together at least a half dozen times over the following decade, combining the group's twin interests in nature and the industrial world. As the roaring twenties approached, Ford increasingly began to worry that his own product, along with the high wages of $5 a day that he had pioneered in 1914, might be drawing people from rural America to the cities, damaging the ability of farmers to live the kind of bucolic, close-to-nature life that Ford cherished from his childhood and continued to idealize. Thus, in 1918 Ford established the village industries program as a way to bring manufacturing jobs to the countryside, allowing residents to reap the economic advantages without giving up their agricultural heritage. His village industries were intended to strengthen rural communities by providing jobs to unemployed and underemployed local residents, allowing farmers to work in the winter and return to farming in the summer.

Ford's village industries also sought to use the original renewable energy source—hydropower. By 1925, nine water-powered plants were built on the sites of abandoned mills, incorporating the original mill structure. Ford was seeking a more sustainable alternative to gasoline. When the manager of the Phoenix facility on the River Rouge, 18.5 miles northwest of today's world headquarters for Ford Motor in Dearborn, Michigan, installed an internal combustion engine to supplement the waterpower, Ford was irate: "We built these plants to run on water power. When I want any other kind of power in, I'll let you know how to do it."[1]

In addition to arguably promoting the E portion of ESG (environmental, social, and corporate governance), Ford also supported the social component as well, although likely for different reasons than advocates of today. Ford matched human and power resources, one man or woman for each horsepower of energy the hydropower plant produced. Ford preferred to hire women for precision manufacturing work. Another of the village industry plants, Green Island, eight miles north of Albany on the Hudson River, eventually employed fifteen hundred people in a town of three thousand.[2]

The village industry program barely outlived Henry since the plants were not financially viable. Strict accounting standards implemented after the OG relinquished the presidency in 1945, at the age of eighty-two, showed that the plants did not return the same cost/benefit ratios as Ford's larger fossil fuel-powered plants. Thus the company began shutting the plants down with only five of the original two dozen surviving into the 1950s and the original Northville plant getting the axe in 1981. Henry had shown willingness to commit and ability to transform the supply network, but his successors did not see an ability to profit.[3]

Fully Committed—Norway's Road to Electrification

Norway believes in electric vehicles—a lot—so much that its official "Visit Norway" proclaims that it is "the EV capital of the world." The site touts that nearing the end of 2023 there were almost six hundred thousand battery EVs in the country and practically two hundred thousand plug-in hybrid EVs, so over 25 percent of the cars on Norwegian roads had some ability to drive on electricity alone.[4] Eighty percent of the vehicles sold in the country in 2022 were electric, thus seemingly on track to reach the political goal of making the entire Norwegian fleet of passenger transport emission free. An example the rest of the world can study? Absolutely. A finished project? Not.

A closer examination reveals that 72 percent of the cars still on roads, over 2 million, are still traditional ICE vehicles. The reader may remember the estimate from chapter 1 that there are almost a billion passenger cars on the planet with an average age of twelve years. Norway has made incredible progress on electrification of transport, yet likely a decade remains before the majority of cars on the roads there are electric. Rather than despairing that the task cannot be accomplished in larger, more car-oriented economies such as the United States, China, and Europe, I argue that we can still learn lessons from the Norwegian experiment by applying the Gears of Change from chapter 2. If done well, the Gears of Change can turn even faster in larger economies to speed up electrification.

Willingness to Commit

While the rest of the world slept, Norway abolished import taxes on zero-emission vehicles beginning in 1990. Next, the country exempted them from other taxes that polluting car owners have to pay while throwing in nice perks, including lower road tolls, free ferry crossings, access to bus lanes, and free public parking. In case you are wondering—there are a lot of ferries in the Scandinavian country—with 127 routes compared to 350 in the United States That means a Norwegian is roughly sixty times more likely to ride on a ferry with their car in a given week, month, or year. Furthermore, battery electric ferries, or BEFs, have already been introduced, with a recent Siemens study estimating that 70 percent of all ferries could be similarly converted.[5]

Following this strong initial step, subsidies to housing associations to facilitate installation of charging stations soon followed. In 2008, Oslo launched the first municipal charging system. By 2015, there were over ten thousand charging stations, roughly one for every five hundred people, children included. Interestingly, Norway has the greatest length of any European country at over one thousand miles north to south. A relatively populated country, ranking 119th in the world, it is also the least densely populated, landing at the 210th spot in world rankings.[6] Norway is far more densely populated than Iceland and Greenland, which are classified as European. Yet, it is virtually a no-persons land compared to the Netherlands, which is over thirty-five hundred times as densely populated. The United States, a very large country by land mass and population, and a country with wide stretches of less populated land, is still roughly two and half times Norway's population density.

To support long-distance trips, a target was set of at least one fast charging station every fifty kilometers on major roads. The first Supercharger was opened in 2016 in Nebbenes, roughly seventy kilometers north of Oslo. This station could accommodate twenty-eight cars at a time. And it was not built by Tesla, although it is the country's bestselling brand, while traditional ICE manufacturers such as Fiat and Renault have seen sales plummet.

Transforming the Supply Network

The chicken-and-egg problem of building cars or charging stations first was solved through subsidies via Enova, the arm of the Norwegian government tasked with funding and advising energy and climate projects in the country. The initial investment of 7 million euros led to nineteen hundred charging stations by 2011. Another significant investment focused specifically on supercharging stations similar to those Tesla built in Europe, the United States, and China. Here, the investment was driven by the government.

Another critical policy decision was to provide financial support to housing associations to install destination chargers. Grants of between 20 and 50 percent of the installation cost were provided in Oslo, Skedmo, Asker, Baerum, and Trondheim. Furthermore, housing associations wanted to install chargers. Why? Because nearly all the electrical power in Norway is renewable and very cheap. The cost per mile to drive on electricity is less than 10 percent of the cost of moving a gasoline-powered car.[7] Today, there are over twenty-five thousand destination charging stations and over six thousand supercharging stations in the country, making the charging process substantially easier than anywhere in the United States. In comparison, California currently has eighty thousand destination and ten thousand superchargers,[8] in a state with nearly 40 million residents. There are 2.5 times as many chargers per capita in Norway than in the Golden State.

Ability to Profit

In a capitalist society, having two of the Gears of Change lubricated and ready to turn is not sufficient. The companies that produce the cars need to be able to profit, and the customers who buy them have to see advantages. The trajectory of the Swedish giant Volvo illustrates the final gear fitting smoothly into place. Selling over six hundred thousand cars worldwide, the company had slightly over $30 billion in revenues in 2022. Annual sales in Norway had hovered around ten thousand units a year from 2010 on, but then, as the transition to electric happened, the Swedish giant began to gain sales. CEO Jim Rowan said the Norwegian combination of incentives and ubiquitous charging "took away all the friction factors."[9]

In 2021, 96 percent of the Volvos sold in the country were Recharge models, with four in ten being BEV and the vast majority of the remainder being plug-in hybrid electric. All in all, the Swedish giant saw its market share grow sales by almost 20 percent year over year from 2020 to 2021.[10] Based on this growth, the decision was announced at the end of 2021 that all Volvo cars sold in Norway would be electric, with no traditional gasoline-only vehicles. The Swedish giant was pleased enough with its profitability and trajectory to be the first company in the world to completely dump conventional gasoline-fueled vehicles in a major market.

Rosy, yet Not Perfect

The shift to electric in Norway has occurred at warp speed relative to most similar large-scale societal and industrial transitions. Several very positive changes have occurred. For one, the air is much cleaner. A monitoring station near Oslo's waterfront collects data on levels of nitrogen oxides, which are produced when gasoline and diesel are the primary fuels. These pollutants cause smog, leading to asthma and other respiratory ailments, which have fallen sharply in lockstep with the rise in EV ownership. Tobias Wolf, Oslo's chief engineer for air quality, claimed in May 2023, "We are on the verge of solving the NO_2 problem."[11] In addition, the amount of electricity demanded from Norway's grid has risen slightly, requiring the Oslo power company Elvia to install additional substations and transformers. Still, there has been no danger of the grid collapsing.

One of the biggest worries in a period of swift societal change is the potential loss of jobs. That threat appears not to have been realized, as even though the pace of change has been fast, many ICE cars still need repairs, so mechanics have steady work. Others have been retrained, including longtime Volkswagen technician Sindre Dranberg, who claims that it was not challenging to retrain to replace defective battery cells in a Volkswagen e-Golf.[12] South of Oslo, a steel plant has become a battery recycling facility. Cirba Industries (profiled in chapter 8) is seeking to do likewise in the United States and Europe. Workers disassemble battery packs, then feed them into a machine that shreds them into four categories: aluminum, copper, plastic, and black mass, which comprises precious ingredients cobalt, manganese, graphite, lithium, and nickel. The factory, owned by Hydrovolt, represents the first of a phased expansion planned throughout the United States and Europe. Currently, the volume of materials available for recycling is small but expected to grow.

There are some challenges, one of which is that waiting lines still occur even with a very substantial investment in charging infrastructure. At a Circle K station and convenience store, roughly one hundred miles south of Oslo, gasoline pumps are in the distinct minority relative to electric chargers. On summer weekends, when city residents travel to summer cottages, the line to recharge can look like the lines seen at US gasoline stations during the oil embargo of the 1970s. Circle K employee Marit Bergsland deals with customers in stride, learning to help customers frustrated by lines connect to chargers in addition to her food preparation and cash register duties. She says, "Sometimes we give them a coffee to calm down."[13] Many apartment dwellers also complain of being unable to charge at convenient times. As examined in chapter 8, matching charger availability with customer demand is a substantial challenge. One solution? At a public hearing on this issue, Oslo's vice mayor, Sirin Hellvin Stav, said that the city's goal is to install further chargers while reducing the number of cars by one-third to enhance safety and walkability.

In his influential book *Hot, Flat and Crowded*, three-time Pulitzer Prize winner Thomas Friedman closed with a chapter titled *China for a Day*. In it, Friedman suggested that if he were the all-powerful leader of America for a single day (with autocratic powers similar to Xi Jinping in China in 2023), he would implement a series of laws and regulations aimed at greening the economy.[14] Then, he would revert to a democratic society with all its messy political battles and debates. In the case of Norway and electric vehicles, we have essentially the opposite of Friedman's thought experiment. This democratic society has made hard decisions and mostly made the Gears of Change turn toward a greener, more environmentally sustainable economy. There are insights the rest of the world can gain from this example.

Predictions for the Companies Profiled: Scoring the Gears of Change

Throughout this book I have examined six companies in some depth. All six have a compelling story and leaders with a vision for a greener, more sustainable and electric future. What do their prospects look like? The following are my predictions for their success over the next half-decade, as depicted in figure 10.1, starting with "best bets" and working toward the longer shots.

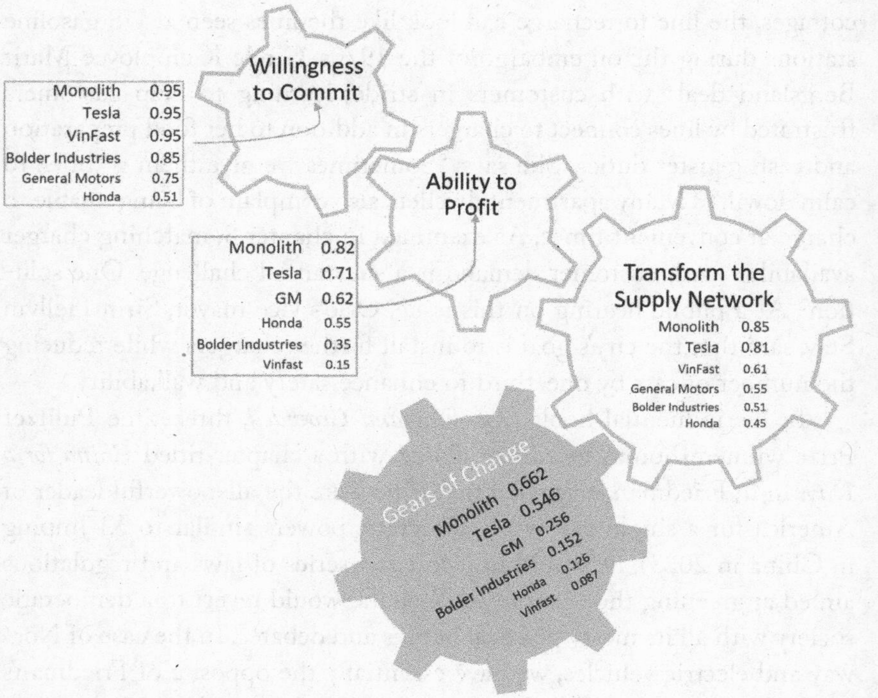

Figure 10.1. Scoring Companies on Gears of Change

Monolith

The definition of a monolith is a large single upright block of stone, especially one shaped into or serving as a pillar or monument, which is a fitting name for a company that I believe will see great financial success. The first cofounders, Rob Hanson and Pete Johnson, spent a decade or more building toward the company now operating in southeast Nebraska, bringing Bill Brady in when needed to help sell into the established industrial community of carbon black and tire producers.

Thus, my estimate of their Willingness to Commit is close to the maximum. Furthermore, the Monolith team has progressed through a carefully orchestrated buildout of its facilities and more extensive supply network in locating in Nebraska and partnering with the Nebraska Public Power District, thus my score of 0.85 on transforming the supply network. Finally, as Bill Brady kept returning to, it's the unit economics that Monolith has built itself around. My estimate of the company's Ability to Profit is high at 0.82. By multiplying the three scores, I arrive at a total Gears of Change score of 0.662, the highest of six companies profiled. Monolith is going to do very well financially over the next decade, I predict, and may well transform the lethargic carbon black industry.

Tesla

Where does one start with this behemoth? Tesla is high on Willingness to Commit, but my prediction is somewhat lower on transforming the supply network. Certainly, Muskla has changed the world, yet now that there is fierce competition, it faces headwinds in making its supercharger network play nice with others. Furthermore, as Alfred Sloan said over a century ago, when saying that GM would make a car for every "purse and purpose," the other automobile manufacturers profiled, plus the many hungry competitors from China and Europe, are coming at the company. The competition for suppliers and customers will be fierce. Tesla may be the reigning champ, but 2023 has shown that it must dynamically adjust prices (typically lower) to match demand to its growing supply of vehicles. Therefore, my score on the Ability to Profit is lower. In total, I land on a Gears of Change score of 0.546, which I believe means Tesla will continue to be a major force in the transportation industry. Still, it will never approach the dominance that GM held in 1960 when it sold over 50 percent of the cars in America.

General Motors

GM was founded by William Durant, who was the largest manufacturer of horse-drawn carriages in the United States in 1900. Durant was initially skeptical of horseless carriages but eventually changed his mind, buying Buick in 1904 and forming the General Motors Company in 1908. One hundred twenty-five years later, the company was led by the inestimable Mary Barra, who also changed her mind and exhibited a very high Willingness to Commit to electrification. GM's Ability to Profit rests on the challenge of balancing the see-saw of traditional demand for ICE vehicles with emerging market for electric, thus a "lower" rating of 0.62. At the same time, Durant's brainchild faces the daunting task of keeping the lights on in its existing gasoline-fueled product line while building confidence and sales in its emerging electric product line, thus a relatively low score for Transforming the Supply Network. In sum, I predict a Gears of Change score of 0.256 for General Motors, suggesting a high chance of success but with substantial risks.

Bolder Industries

Much like Monolith, Bolder has been led by a visionary founder, Tony Wibbeler, who came to the venture with an extraordinary Willingness to Commit. Tony and the Bolder team have followed a phased expansion that mirrors Monolith's with a trial facility in Maryville and a larger facility in Terre Haute. Unfortunately for Bolder, the high score on Willingness to Commit runs into a lower score for Transforming the Supply Network. The challenges associated with collecting and processing used tires are immense.

Simply put, creating a circular economy for any product is extraordinarily difficult. Bolder has assembled a supply partnership, however, that puts it into the game, thus my positive score of 0.51. Finally, the Ability to Profit is the most challenging gear for Bolder. The investors and trading partners have bought into the economic forecasts of the leadership team, yet there is little margin for error. My estimate of a 0.35 score for the Ability to Profit results in a total Gears of Change score of 0.152.

Honda

Soichiro's brainchild has been late to the table and thus receives a reasonably low Willingness to Commit score. The nervous giant has been slow to begin supply network transformation and is forced to play catchup. At the same time, it has formed key partnerships with GM and Sony, the latter providing a refreshing view of cars. Announced in January 2023, Sony and Honda unveiled an alliance in a new brand, Afeela, in which the pair of Japanese corporate giants planned to offer an EV beginning in 2026 that would be leased for a ten-year life span. Sony and Honda believe that the hardware portions of cars will last ten years as customers can update the cars via software updates to keep the cars fresh. According to Sony Honda Mobility CEO Yasuhide Mizuno, "Replacing the car every three to five years is a very traditional methodology. But now: big change. This car is always updating; therefore, we try to utilize customers over five to ten years."[15] Finally, Honda is quite profitable today. Thus the company has a path to titrate profitability from gasoline cars to support the transition to EV, therefore my rating of 0.55 for the Ability to Profit. Honda's prospects do not look as good as Tesla or Monolith, but it would be not very smart to bet against Soichiro's brainchild.

VinFast

By force of personality, Pham Nat Vuoung has become the wealthiest man in Vietnam, starting with noodle production in the Ukraine. Vuong clearly is willing to commit with the audaciousness to make twin bold decisions, first to build an automotive manufacturer in the first place, then to forgo the production of gasoline cars to race down the EV learning curve. Thus, a Willingness to Commit rating of 0.95. My estimate for Transforming the Supply Network is a bit higher than Honda's because VinFast does not have nearly the same size network of investments in the production of gasoline vehicles. Vuong decided to build a battery production plant east of Hanoi several years before Honda made a similar commitment. VinFast's Achilles' heel is the Ability to Profit. They were having to largely self-finance the company; at some point even a billionaire's pockets empty. Throughout August 2023 the company became a focus of broad interest as the stock debuted on the NASDAQ through a backdoor maneuver, listing with Black Spade as a special purpose acquisition company. VinFast's nominal market value soared to over $200 billion on August 28, 2023. This was for a company burning through cash at a rate of $3 to $5 billion a year. Four days later the stock closed at a share price that valued the company under $70 billion. And even this number is too high. First, less than 1 percent of the company was available for trading—thus, the initial hype was people essentially gambling. Second, as Zachary Abuza wrote in the *South China Morning Post*, "They only got this far by cutting corners, which might be possible in the crony-capital world of Vietnam, but it won't work in the United States."[16]

Painful but also not wrong. My prediction is that the company has achieved some fantastic things, simply building an automotive plant that can produce vehicles at scale in Vietnam and be competitive with existing significant brands is a Herculean feat. The decision to switch from ICE to EV and go all in was perilous, as it deprived the company of a means of selling a higher volume of cars and financing the build-out to EV.

However, life is filled with risk, and the decision also allowed the company to commit to electric and build a battery production facility long before Honda did. That battery plant and its associated people have undoubtedly learned quite a bit about battery production. As I write this, the day before Labor Day 2023, the United Auto Workers and the Big 3 of GM, Ford, and Stellantis are headed toward a strike in mid-September. Chinese manufacturers are looking stronger and stronger in the EV ecosystem. And VinFast is listed in America and paddling like a duck swimming—working furiously under the surface of the water yet barely moving on top of it. I believe Vuong is negotiating with potential partners and buyers to sell a large stake in the company. An infusion of $5 to $10 billion in capital would breathe new life into VinFast.

Time will tell. I appreciate your time and attention, dear reader. This is indeed an American revolution, but Americans should never forget that it is really a World revolution. Many people and companies are fighting to grab market share. Jim DeLuca, a prince of the Motor City, is in Saudi Arabia seeking to build Ceer into a significant player. And Mohamed Bin Salman and the Saudis have some of the deepest pockets in the world.

Notes

Preface

1. "The History of Self-Fueling," NACS, April 5, 2023, https://www.convenience.org/Topics/Fuels/The-History-of-Self-Fueling.

Chapter 1

1. Alexi Friedman, "Former GM Plant Is Razed for Linden Renaissance," *Star-Ledger* (Newark), October 31, 2023).

2. Shelby L. Stanton, *The Rise and Fall of an American Army: U.S. Ground Forces in Vietnam, 1963–1973* (New York: Random House, 2007), 366–68.

3. Guenter Lewy, *America in Vietnam* (New York: Oxford University Press, 1978).

4. John Reed, "The Rise and Rise of a Vietnamese Corporate Empire," *Financial Times*, June 26, 2019, https://www.ft.com/content/84323c32-9799-11e9-9573-ee5cbb98ed36.

5. "Greenhouse Gas Emissions from a Typical Passenger Vehicle," United States Environmental Protection Agency, August 28, 2023, https://www.epa.gov/greenvehicles/greenhouse-gas-emissions-typical-passenger-vehicle#typical-passenger.

6. "Volkswagen Questioned by U.S. Senator over Human Rights Issues Linked to China Suppliers," *Automotive News*, April 28, 2022, https://www.autonews.com/china/volkswagen-questioned-us-sen-marco-rubio-over-human-rights-issues-linked-china-suppliers.

7. Yasmena AlMulla, "Kuwait: Fire Ripped through 25,000 Square Meters of Al Sulabiya Tire Site," *Gulf News*, October 23, 2020, https://gulfnews.com/world/gulf/kuwait/kuwait-fire-ripped-through-25000-square-meters-of-al-sulabiya-tire-site-1.74769506.

8. John Arlidge "Mary the Mechanic Pulls GM out of the Wreckage," *Sunday Times* (London), September 20, 2015.

9. Mark Matousek, "Mary Barra Was Called a 'Lightweight' When She Became CEO of GM—Here's How She Transformed the Company and Silenced Her Doubters," *Business Insider*, January 11, 2018, https://www.businessinsider.com/heres-how-mary-barra-silenced-critics-who-called-her-a-lightweight-2018-1.

10. Alisa Priddle, "GM Names Clegg New Production Chief for N. America," *Detroit Free Press*, June 24, 2014.

11. Ford Motor Co., "*1909–1927 Ford Model T, TT Service Manual*" (Detroit: Detroit Iron, 1927).

12. Douglas Brinkley, "Prime Mover," *American Heritage*, June 2003.

13. Joe McDonald, "China Auto Show Highlights Intense Electric Car Competition," Associated Press, April 15, 2023, https://apnews.com/article/china-auto-electric-vehicle-0ea339b7be33090b222169fd433cf51c.

14. Adam Selzer, "Burnham's 'Make No Little Plans' Quote: Apocryphal No More!," Mysterious Chicago, March 3, 2019, https://mysteriouschicago.com/finding-daniel-burnhams-no-little-plans-quote.

15. Elana Scherr, "VinFast Is Going All-Electric, and Worldwide, at Full Speed," *Car and Driver*, April 27, 2022, https://www.caranddriver.com/features/a39816283/vinfast-update-factory-tour-vf8-prototype-drive/.

16. Thanh Nhan, "Vietnam Auto Newbie VinFast Zooms to Fifth Place in Sales," *VnExpress International*, April 26, 2020, https://e.vnexpress.net/news/business/companies/auto-newbie-vinfast-zooms-to-fifth-place-in-sales-4090698.html.

17. Jack Ewing, "U.S. Electric Car Sales Climb Sharply Despite Shortages," *New York Times*, July 14, 2022, https://www.nytimes.com/2022/07/14/business/electric-car-sales.html?referringSource=articleShare.

18. "U.S. Automobile Production Figures," Wikipedia, accessed July 6, 2023, https://en.wikipedia.org/wiki/U.S._Automobile_Production_Figures.

19. Reed, "The Rise and Rise of a Vietnamese Corporate Empire."

20. "Sun Group (Vietnam)," Wikipedia, accessed July 8, 2023, https://en.wikipedia.org/wiki/Sun_Group_(Vietnam).

21. Matt Talhelm, "Look Inside VinFast's Vietnam SUV Plant and How It Compares to the One Coming to NC," WRAL *TechWire*, July 21, 2022, https://wraltechwire.com/2022/07/21/look-inside-vinfasts-vietnam-suv-plant-and-how-it-compares-to-the-one-coming-to-nc/.

22. MikeL01, "VinFast to Stop Making Gas-Powered Cars by End-August," VinFast Talk, July 16, 2022, https://www.vinfasttalk.com/threads/vinfast-to-stop-making-gas-powered-cars-by-end-august.127/.

23. "Statement by President Biden on Electric Vehicle and Battery Manufacturing Investments in North Carolina," Statements and Releases, The White House, March 29, 2022, https://www.whitehouse.gov/briefing-room/statements-releases/2022/03/29/statement-by-president-biden-on-electric-vehicle-and-battery-manufacturing-investments-in-north-carolina/.

24. Anshuman Daga and Phuong Nguyen, "Vietnam's VinFast Taps Banks for $4 Bln EV Factory Funding Deal," Reuters, July 12, 2022, https://www.reuters.com/business/autos-transportation/vietnams-vinfast-says-agrees-4-bln-ev-factory-funding-with-credit-suisse-citi-2022-07-13/.

25. Nguyen Kieu Giang, "Vietnamese EV Maker VinFast Files for US IPO," *Bloomberg*, December 6, 2022, https://www.bloomberg.com/news/articles/2022-12-06/vietnamese-electric-vehicle-maker-vinfast-files-for-us-ipo.

Chapter 2

1. Information on the development of automatic transmissions and the differences in gear usage between ICEs and EVs is available at https://www.mistertransmission.com/a-brief-history-of-the-automatic-transmission/ and https://www.drivingelectric.com/your-questions-answered/95/do-electric-cars-have-gearboxes.

2. Zeyi Yang, "How did China Come to Dominate the World of Electric Cars," *MIT Technology Review*, February 21, 2023, https://www.technologyreview.com/2023/02/21/1068880/how-did-china-dominate-electric-cars-policy/

3. Andrea Lenschow, Carina Sprungk, "The Myth of a Green Europe," *Journal of Common Market Studies* 48, no. 1 (2009): 133–54, https://doi.org/10.1111/j.1468–5965.2009.02045.x.

4. Henning Deters, "European Environmental Policy at 50: Five Decades of Escaping Decision Traps," *Environmental Policy and Governance* 29 (May 17, 2019): 315–25.

5. Dan Mihalascu, "Ram CEO: High Electrification Costs Are 'Elephant in the Room': Stellantis." Inside EVs, April 10, 2023. https://insideevs.com/news/661549/high-electrification-costs-elephant-in-room-stellantis/amp.

6. Elizabeth Partsch, "Europe's Transition to Electric Vehicles: How It's Going, and What Lies Ahead," *Impakter*, December 24, 2022, https://impakter.com/europes-transition-to-electric-vehicles-hows-it-going.

7. Yang, "How Did China Come to Dominate?"

8. "Editorial: EV Battery Boom: Manchin Gambit Might Be Working," *Automotive News*, October 16, 2022, https://www.autonews.com/editorial/ev-battery-boom-joe-manchins-gambit-might-be-working.

9. "Inflation Reduction Act of 2022," United States Department of Energy, July 14, 2023, https://www.energy.gov/lpo/inflation-reduction-act-2022.

10. Loukas, "EV vs ICE: Similarities and Differences," ArenaEV, April 24, 2022, https://www.arenaev.com/ice_v_ev__differences_and_similarities-news-185.php.

11. Tim Levin, "These Are the 20 Car Brands with the Most Loyal Customers," *Business Insider*, July 17, 2020, https://www.businessinsider.com/car-buying-brands-most-loyal-customers-automotive-sales-loyalty-subaru-2020-7#1-subaru-20.

12. Clive Thompson, "When Coal First Arrived, Americans Said 'No Thanks,'" *Smithsonian*, July–August 2022, https://www.smithsonianmag.com/innovation/americans-hated-coal-180980342.

13. Thompson, "When Coal First Arrived," 107.

14. Michael Wayland, "Why You Won't See Many Car Ads during Sunday's Super Bowl," CNBC, February 15, 2023, https://www.cnbc.com/2023/02/10/gm-jeep-kia-super-bowl-ads.html.

15. Brad Plumer and Lisa Friedman, "A Swaggering Clean-Energy Pioneer, with $400 Billion to Hand Out," *New York Times*, May 11, 2023, https://www.nytimes.com/2023/05/11/climate/jigar-shah-climate-biden.html.

16. Data gathered from Mergent Online, which aggregates data from thousands of companies on customers and suppliers.

17. Benchmark Mineral Intelligence, *Ohio Battery Supply Chain Opportunities*, JobsOhio, February 2022, https://cdn.bfldr.com/AHJE351Z/at/cx67vf3tm69g4r49w5v3z/BMI_Ohio_Report.pdf.

18. Mark Williams, "Honda Battery Plant Helps Ohio Maintain Edge in Changing Auto Industry," *Columbus Dispatch*, October 16, 2022.

19. John Irwin, David Kennedy, "Magna Pegs $700M for EV Battery Enclosures," *Automotive News*, February 18, 2023, https://www.autonews.com/suppliers/magna-spend-michigan-ontario-bid-be-big-ev-player.

20. "Kaiser-Frazer," Wikipedia, https://en.wikipedia.org/wiki/Kaiser-Frazer (July 16, 2023).

21. "Edsel," Wikipedia, https://en.wikipedia.org/wiki/Edsel (July 16, 2023).

22. "How GM Destroyed Its Saturn Success," *Forbes*, March 10, 2010, https://www.forbes.com/2010/03/08/saturn-gm-innovation-leadership-managing-failure.html?sh=68cbd5e46ee3.

23. Morgon Korn, "Hummer Electric Pickup Truck Unveiled: What You Need to Know," ABC News, October 20, 2020, https://abcnews.go.com/Business/hummer-electric-pickup-truck-unveiled/story?id=73693774.

24. Mitchell Clark, *The Verge*, June 30, 2022, https://www.theverge.com/.

25 . "John DeLorean," Wikipedia, https://en.wikipedia.org/wiki/John_DeLorean (July 14, 2023).

26. Kat Bailey, "The DeLorean Is Officially Back, and It's Electric," *IGN Southeast Asia*, February 15, 2022, https://sea.ign.com/back-to-the-future-theater/181994/news/the-delorean-is-officially-back-and-its-electric.

27. "VinFast Delays US Electric Vehicle Plant Operation to 2025," Reuters, March 10, 2023, https://www.reuters.com/business/autos-transportation/vinfast-delays-us-electric-vehicle-plant-operation-2025-2023-03-10/.

28. "Honda," Wikipedia, https://en.wikipedia.org/wiki/Honda (July 11, 2023).

29. "Honda's Marysville Auto Plant at 25 Years," Honda, August 21, 2007, https://hondanews.com/en-US/releases/release-4a1b74211ee992e8cc546d004c34be69-hondas-marysville-auto-plant-at-25-years-historic-yet-new.

30. Henry Mintzberg, Richard T. Pascale, Michael Goold, Richard Rumelt, "The Honda Effect Revisited," *California Management Review* 34, no. 1 (Spring, 1996): 78–91.

31. C. K. Prahalad and Gary Hamel, "The Core Competence of the Corporation," *Harvard Business Review*, May–June 1990, https://hbr.org/1990/05/the-core-competence-of-the-corporation.

32. Andrew Hawkins, "GM and Honda Expanding Their Partnership to Develop 'Millions' of Affordable EVs in 2027," *The Verge*, April 5, 2022, https://www.theverge.com/2022/4/5/23011201/gm-honda-ev-partnership-millions-affordable-2027.

Chapter 3

1. Martin LaMonica, "Tesla Motors Founder: Now There Are Five," *CNET*, September 21, 2009. https://www.cnet.com/culture/tesla-motors-founders-now-there-are-five/.

2. Ashlee Vance, *Elon Musk: Tesla, SpaceX and the Quest for a Fantastic Future* (New York: HarperCollins, 2015).

3. A. J. Caldwell, "Why Top Automakers Spend Millions on Concept Cars They Don't Plan on Making," *Business Insider*, January 27, 2021, https://www.businessinsider.com/automakers-spend-millions-on-concept-cars-they-dont-make-2019-4.

4. David Thomas, "Checking Out the Tesla Electric Roadster," https://www.cars.com/articles/checking-out-the-tesla-electric-roadster-1420663326614/, December 1, 2006.

5. "Tesla Logo: Meaning, PNG, and Transparent Logo," *Car Logos.org*, May 19, 2023, https://www.carlogos.org/car-brands/tesla-logo.html. Andrew J. Hawkins, "GM Unveils New Logo to Emphasize Its Pivot to Electric Vehicles," *The Verge*, January 8, 2021.

6. Garrett Mast, "Toyota to Collaborate with Redwood Materials on a Sustainable, Closed-Loop Electrified Vehicle Battery Ecosystem, June 21, 2022, https://pressroom.toyota.com/toyota-to-collaborate-with-redwood-materials-on-a-sustainable-closed-loop-electrified-vehicle-battery-ecosystem/.

7. Tim Keenan, "American Battery Solutions Acquires Tiveni's Entire Battery Systems IP Portfolio," *DBusiness*, September 14, 2022, https://www.dbusiness.com/daily-news/american-battery-solutions-acquires-tivenis-entire-battery-systems-ip-portfolio/.

8. Neal E. Boudette, "For Tesla, 'Production Hell' Looks Like the Reality of the Car Business," *New York Times*, April 3, 2018, https://www.nytimes.com/2018/04/03/business/tesla-model-3.html.

9. Boudette.

10. Tim Lasseter, Patrick W. Houston, Joshua L. Wright, and Juliana Y. Park, "Amazon Your Industry: Extracting Value from the Value Chain," *Strategy & Business*, January 1, 2000, https://www.strategy-business.com/article/10479.

11. Erik Brynjolfsson and Michael D. Smith, "Frictionless Commerce? A Comparison of Internet and Conventional Retailers," *Management Science* 46, no. 4 (April 2000): 563–85.

12. Evan Lasseter, "Average Prices for New EVs Are Falling: 'Still a Luxury' for Most, but That's Changing as More Are Produced," *Atlanta Journal-Constitution*, June 18, 2023.

13. Shannon Osaka, "Elon Musk Agrees to Open Parts of Tesla's Charging Network to Everyone," *Washington Post*, February 15, 2023, https://www.washingtonpost.com/climate-environment/2023/02/15/tesla-supercharger-network-locked/#.

14. Fred Lambert, "Tesla Tries to Get Owners To Give Up 'Unlimited Free Supercharging for Life," *Electrek*, April 24, 2023, https://electrek.co/2023/04/24/tesla-triesowners-give-up-unlimited-free-supercharging-for-life/.

15. Robert Ferris, "Tesla Will Need Billions to Make US Supercharger Network Compete with Gas Stations, Says Analyst," CNBC, March 3, 2017, https://www.cnbc.com/2017/03/03/tesla-needs-billions-to-make-supercharger-network-rival-gas-stations.html.

16. Loren McDonald, "UBS Analyst Gets Future Investment Costs for Tesla Supercharger Network Super Wrong," CleanTechnica, March 4, 2017, https://cleantechnica.com/2017/03/05/ubs-analyst-gets-future-investment-costs-tesla-supercharger-network-super-wrong/.

17. Jack Ewing, "Tesla May Already Have Won the Charging Wars," *New York Times*, June 27, 2023, https://www.nytimes.com/2023/06/27/business/energy-environment/tesla-gm-ford-charging-electric-vehicles.html.

18. Robert Cyran, "The iPad Factor in Digital Readers," *New York Times*, June 28, 2010, https://www.nytimes.com/2010/06/28/technology/28views.html.

19. Ewing, "Tesla May Already Have Won."

20. "Little's Law," Wikipedia, December 18, 2023, https://en.wikipedia.org/wiki/Little%27s_law.

21. David Maister, "The Psychology of Waiting Lines," *David Maister*, 1985, https://davidmaister.com/articles/the-psychology-of-waiting-lines/.

22. Ellen Edmonds, "Nearly 51 Million Americans to Travel This Thanksgiving, Highest Volume in a Dozen Years," AAA Newsroom, November 16, 2017, https://newsroom.aaa.com/2017/11/nearly-51-million-americans-travel-thanksgiving-highest-volume-dozen-years/.

23. "Jealousy Quotes," Goodreads, https://www.goodreads.com/quotes/tag/jealousy.

24. Christopher Cox, "Elon's Appetite for Destruction," *New York Times*, January 17, 2023, https://www.nytimes.com/2023/01/17/magazine/tesla-autopilot-self-driving-elon-musk.html.

Chapter 4

1. Jim DeLuca, personal interview with author, June 21, 2023.

2. Paul Lienert, "New GM CEO Barra Seen as a Winner Early in Her Career," Reuters, December 10, 2013, https://www.reuters.com/article/uk-autos-barra-personal/new-gm-ceo-barra-seen-as-a-winner-early-in-her-career-idUKBRE9BA00H20131211.

3. Donald W. Nauss, "GM's Man Who Bested NBC Helps Rouse Sleeping Giant," *Los Angeles Times*, February17, 1993, https://www.latimes.com/archives/la-xpm-1993-02-17-mn-238-story.html.

4. Ben Klayman, "GM Plans 500,000 Vehicles with Electrification by 2017," Reuters, November 14, 2012, https://www.reuters.com/article/us-gm-evs/gm-aims-to-build-500000-electric-technology-vehicles-a-year-idUKBRE8AD1EM20121114.

5. James Amend, "Mary Barra Appointment Puts Insider Back in GM CEOs Seat," *Ward's Auto*, December 1, 2013, https://www.wardsauto.com/industry/mary-barra-appointment-puts-insider-back-gm-ceo-s-seat.

6. Siobhan Hughes, "Senators Challenge GM's Barra, Push for Faster Change," Wall Street Journal, April 2, 2014.

7. Jim DeLuca, personal interview with author, June 21, 2023.

8. Manoli Katakis, "Executive Vice President of General Motors Global Manufacturing Tim Lee to Retire April 1," GM Authority, January 26, 2014, https://gmauthority.com/blog/2014/01/executive-vice-president-of-general-motors-global-manufacturing-tim-lee-to-retire-april-1/.

Chapter 5

1. Personal interview with author, April 7, 2023.

2. John Reed, "GM Forms Partnership with Vietnam's VinFast," *Financial Times*, June 28, 2018, https://www.ft.com/content/a8da1eb4-7aa9-11e8-bc55-50daf11b720d.

3. Jim DeLuca, personal interview with author, June 21, 2023.

4. Alissa Priddle, "VinFast Is Making Carmaking 101 Look Easy," *Motor Trend*, April 29, 2022, https://www.motortrend.com/news/vinfast-vietnam-factory-tour/.

5. Phuong Nguyen, "Vietnam's VinFast Ships First Electric Vehicles to U.S. Customers," Reuters, November 25, 2022, https://www.reuters.com/business/autos-transportation/vietnams-vinfast-ships-first-electric-vehicles-us-customers-2022-11-25/.

6. "VinFast Officially Delivers First VF 8 City Edition Vehicles to U.S. Customers," VinFast, March 1, 2023.

7. Donut, "We Drove the Worst Reviewed Car in America," YouTube, July 12, 2023, https://www.youtube.com/watch?v=DF7kaLTsNHQ.

8. Phuong Nguyen, "Vietnam's VinFast recalls first batch of US-bound EVs Over Safety Risk," Reuters, May 25, 2023, https://www.reuters.com/business/autos-transportation/vietnams-vinfast-recalls-all-first-batch-evs-shipped-us-over-security-risk-2023-05-25/.

9. Luisa Beltran, "VinFast, the Latest EV IPO, Comes to U.S. from Vietnam," *Barron's*, April 7, 2022, https://www.barrons.com/articles/vinfast-ev-ipo-vietnam-51649367803.

10. Joe White, "VinFast: Party Like Its 2021," Reuters, December 7, 2022.

11. Phuong Nguyen, "VinFast Posts Deeper Loss in Q1, Eyes July Completion of SPAC Merger," Reuters, June 16, 2023, https://www.reuters.com/technology/vinfast-posts-deeper-loss-q1-eyes-july-completion-spac-merger-2023-06-16/.

12. Sean O'Kane, "Lucid Motors Goes Public, Collects $4.5 Billion," *The Verge*, July 26, 2021, https://www.theverge.com/2021/7/26/22594177/lucid-motors-spac-nasdaq-saudi-arabia-ev-startup.

13 . "2024 Lucid Air," *Car and Driver*, August 8, 20023, https://www.caranddriver.com/lucid-motors/air.

14. John Rosevear, "Lucid Misses Revenue Expectations After EV Deliveries Disappoint," CNBC, August 7, 2023, https://www.cnbc.com/2023/08/07/lucid-lcid-q2-2023-earnings-guidance.html.

15. "Black Spade Acquisition Co Announces Extension of Combination Period," *PR Newswire*, July 14, 2023, https://www.prnewswire.com/news-releases/black-spade-acquisition-co-announces-extension-of-combination-period-301877405.html?tc=eml_cleartime.

16. Jim DeLuca, personal interview with author, June 21, 2023.

17. Personal interview with author, July 11, 2023.

18. Yuji Nitta, "Vietnam Vingroup's Ambition to Take on Tesla Hits Bumpy Road," *Nikkei Asia*, August 3, 2023, https://asia.nikkei.com/Business/Automobiles/Vietnam-Vingroup-s-ambition-to-take-on-Tesla-hits-bumpy-road.

19. Brian Gordon and Tyler Dukes, "VinFast Lobbied Gov. Cooper to Advance Federal Loan Bid to Avoid 'Further' NC Delays," *News & Observer*, August 8, 2023, https://www.newsobserver.com/news/politics-government/article277702563.html.

20. "'Crown Jewel': Electric Carmaker VinFast Breaks Ground in Its $4B NC Manufacturing Plant," *ABC 11 News*, July 28, 2023, https://abc11.com/VinFast-north-carolina-chatham-county-roy-cooper-electric-vehicle/13562386/.

21. Andrew Lambrecht, "VinFast and Black Spade Announce $27B Merger for August IPO," *InsideEVs*, July 31, 2023, https://insideevs.com/news/679514/VinFast-secured-by-sec/.

22. Richard Stradling, "Photographer Who Took Iconic Vietnam War Photo Shoots Pics at VinFast Groundbreaking, *Raleigh News & Observer*, July 28, 2023, https://www.newsobserver.com/news/business/article277745313.html.

23. Anuruddha Ghosal, "Vietnam's VinFast Committed to Selling EVs to US Despite Challenges, Intense Competition," *ABC News*, October 24, 2023, https://abcnews.go.com/International/wireStory/vietnams-vinfast-committed-selling-evs-us-despite-challenges-104269388.

24. Jaelyn Campbell, "VinFast and Yorkville Reach $1 Billion Share Deal," CBT News, October 20, 2023, https://www.cbtnews.com/vinfast-and-yorkville-reach-1-billion-share-deal/.

25. Laurence Iliff, "VinFast Considers Ultralow-Priced VF 3 EV for U.S. at Urging of Prospective Dealers." *Automotive News*, October 27, 2023, https://www.autonews.com/retail/vinfast-gauges-dealer-interest-sub-20000-ev.

Chapter 6

1. "GM CEO Mary Barra Talks Sustainability with Chelsea Clinton during Climate Week," *3BL*, September 26, 2014, https://www.3blmedia.com/news/gm-ceo-mary-barra-talks-sustainability-chelsea-clinton-during-climate-week.

2. Jim DeLuca, personal interview with author, June 21, 2023.

3. Jim DeLuca, personal interview with author, June 21, 2023.

4. "GM Reports Record Net Income of $9.7 Billion and Record EBIT-Adjusted of $10.8 Billion for 2015," General Motors Newsroom, accessed August 7, 2023, https://news.gm.com/newsroom.detail.html/Pages/news/emergency_news/2016/0203-2015-4th-qtr-earnings.html.

5. "Spark EV," Wikipedia, https://en.wikipedia.org/wiki/Chevrolet_Spark#Spark_EV

6. Joseph White, "GM Buys Cruise Automation to Speed Self-Driving Car Strategy," Reuters, March 11, 2016, https://www.reuters.com/article/us-gm-cruiseautomation-idUSKCN0WD1ND.

7. Brian Fung, "The Chevy Bolt Looks to Dethrone Tesla. Here Are Our First Impressions," *Washington Post*, September 21, 2016, https://www.washingtonpost.com/news/the-switch/wp/2016/09/21/the-chevy-bolt-aims-to-dethrone-tesla-here-are-our-first-impressions/.

8. Lindsay Vanhulle, "GMC to Keep Hummer EV Customers Engaged as SUV Sales Begin," *Automotive News*, March 23, 2023, https://www.autonews.com/cars-concepts/gmc-hummer-ev-demand-growing-even-while-reservations-hold.

9. Mark Kane, "GMC Hummer EV Pickup Sales Down 83 Percent In Q2 2023," *InsideEVs*, July 5, 2023, https://insideevs.com/news/675394/us-gmc-hummer-ev-sales-2023q2/.

10. Kris Holt, "GM Modernizes Its Logo to Highlight Its EV-Centric Future," Engadget, January 8, 2021, https://www.engadget.com/gm-logo-branding-electric-vehicles-210607845.html.

11. Jamie LaReau, "GM Forms Alliance with Honda to Develop Future Products in North America," *Detroit Free Press*, September 3, 2020, https://www.freep.com/

story/money/cars/general-motors/2020/09/03/general-motors-honda-partnership-north-america/5701104002/#.

12. "GM Builds Electric Battery Lab in Michigan as It Tries to Cut EV Costs, Extend Range," *Washington Post*, October 5, 2021, https://www.washingtonpost.com/business/economy/gm-builds-electric-battery-lab-in-michigan-as-it-tries-to-cut-ev-costs-extend-range/2021/10/05/320d5186-25cb-11ec-8831-a31e7b3de188_story.html.

13. Mike Ramsey and Gautham Nagesh, "LG Electronics to Be Key Supplier for GM Bolt," *Wall Street Journal*, October 20, 2015, https://www.wsj.com/articles/lg-electronics-to-be-key-supplier-for-gm-bolt-1445377888.

14. Mike Colias, "GM to Recover $1.9 Billion in Bolt-Recall Costs in Deal With LG," *Wall Street Journal*, October 12, 2021, https://www.wsj.com/articles/gm-says-it-will-recover-1-9-billion-of-bolt-recall-costs-from-lg-11634042381.

15. Claire Bushey, "GM Tries to Catch Tesla by Following Its Supply Chain Playbook," *Financial Times*, June 18, 2023, https://www.ft.com/content/8e9b17d3-5e1c-4aad-9e28-e6b96648dd7b.

16. Alan Ohnsman, "California's Lithium Rush for EV Batteries Hinges on Taming Toxic, Volcanic Brine," *Forbes*, August 31, 2022, https://www.forbes.com/sites/alanohnsman/2022/08/31/californias-lithium-rush-electric-vehicles-salton-sea/?sh=6358828a4f63.

17. Ernest Schneyder, "GM to Help Lithium Americas Develop Nevada's Thacker Pass Mine," Reuters, January 31, 2023, https://www.reuters.com/markets/commodities/gm-lithium-americas-develop-thacker-pass-mine-nevada-2023-01-31/.

18. Bushey, "GM Tries to Catch Tesla."

19. Scott Sonner, "9th Circuit Denies Bid by Environmentalists and Tribes to Block Nevada Lithium Mine," Associated Press, July 17, 2023, https://apnews.com/article/nevada-thacker-pass-lithium-mine-4ad772a6940eb8edd507b50a179202f2.

20. Jennifer Solis, Court Asked to Vacate NV Lithium Mine Approval after Precedent-Setting Case, *Nevada Current*, June 28, 2023, https://www.nevadacurrent.com/2023/06/28/court-asked-to-vacate-nv-lithium-mine-approval-after-precedent-setting-case/.

21. Wei Wen, "Cadillac's First LYRIQ EV Rolls Off Assembly Line in Shanghai," *Yicai Global*, May 9, 2022, https://www.yicaiglobal.com/news/cadillac-first-lyriq-ev-rolls-off-assembly-line-in-shanghai.

22. Robinson Meyer, "America Can't Build a Green Economy without China," *New York Times*, July 17, 2023, https://www.nytimes.com/2023/07/17/opinion/america-china-clean-energy.html.

23. Matthias Holweg, "The Genealogy of Lean Production," *Journal of Operations Management* 25, no. 2 (March 2007): 420–37, https://doi.org/10.1016/j.jom.2006.04.001.

24. Keith Williams, "'I'm Out': Ford's F-150 Lightning Price Hikes Are Costing It Customers," *The Verge*, June 27, 2023, https://www.theverge.com/2023/6/27/23771407/ford-f150-lightning-price-increase-cancel-order-dealer.

25. Scooter Doll, "Ford F-150 Lightning gets $10k Price Cut," *Electrek*, July 17, 2023, https://electrek.co/2023/07/17/ford-f-150-lightning-production-scale-price-cut-near-10000-new-customers/.

26. "General Motors Takes Q3 Sales Crown, but GM Stock Makes New Low," *Investor's Business Daily*, October 4, 2023, https://www.investors.com/news/auto-sales-q3-2023-gm-ford-gains/.

27. Michael Wayland, "Why GM Is Killing the Chevy Bolt—America's Cheapest EV—Amid Record Sales," CNBC, April 29, 2023, https://www.cnbc.com/2023/04/29/why-gm-is-killing-the-chevy-bolt-ev-amid-record-sales.html.

28. Wayland.

29. Joseph White and David Shepardson, "UAW Reaches Deal with GM, Ending Strike against Detroit Automakers," Reuters, October 30, 2023, https://www.reuters.com/business/autos-transportation/gm-reaches-tentative-deal-with-uaw-source-says-2023-10-30/.

30. "Auto Strike Settlements Will Raise Costs for Detroit's Big 3," *Business Inquirer*, November 1, 2023, https://business.inquirer.net/429598/auto-strike-settlements-will-raise-costs-for-detroits-big-3.

31. Hannah Lutz, "EV Transition Slows as Inventory Grows and Industry Hits Hurdles," *Automotive News*, October 31, 2023, https://www.autonews.com/retail/ev-transition-slows-amid-inventory-woes-sluggish-demand.

Chapter 7

1. Andi Gates, "Supermarket Sells Eggs and Equities," *Columbus Dispatch*, 27 March 1983, 10.

2. Duane St. Clair, "State Wrapping Up Honda Negotiations," *Columbus Dispatch*, 27 September 1977, 1

3. Duane St. Clair, "Honda Site Report Starts State Probe," *Columbus Dispatch*, September 28, 1977, 1–3

4. "Honda Makes Major Investment in Ohio to Create New Electric Vehicle Hub," Honda Media Newsroom, October 11, 2022, https://hondanews.com/en-US/releases/release-1503019bd8a757ea08267d79443a4c43-honda-makes-major-investment-in-ohio-to-create-new-electric-vehicle-hub.

5. "Governor DeWine Announces Honda to Invest in Ohio for Electric Vehicle Production, Including New Battery Plant with LG Energy Solution," Office of the Governor of Ohio, press release, October 11, 2022, https://governor.ohio.gov/media/news-and-media/governor-dewine-announces-honda-to-invest-in-ohio-for-electric-vehicle-production-including-new-battery-plant-with-lg-energy-solution-10112022.

6. "Mobility at the Ohio State University," The Ohio State University, accessed September 11, 2023, https://www.osu.edu/research/mobility/.

7. Dr. Giorgio Rizzoni, personal interview with author, July 27, 2023.

8. KBB Editors, "Honda Hybrid Sales Reach 1 Million Worldwide," *Kelly Blue Book*, October 16, 2012, https://www.kbb.com/car-news/honda-hybrid-sales-reach-1-million-worldwide/.

9. Hans Greimel, "Honda CEO Toshihiro Mibe Outlines Multistep Plan to Jump Ahead in EVs," *Automotive News*, April 26, 2023, https://www.autonews.com/automakers-suppliers/honda-ceo-pledges-fight-back-behind-ev-race.

10. Mark Williams, "Meeting the Challenge: Ohio Report Outlines the Challenge Ahead in Move to EVs," *Columbus Dispatch*, July 24, 2023.

11. "About NSF Engines," National Science Foundation, accessed August 14, 2023, https://new.nsf.gov/funding/initiatives/regional-innovation-engines/about-nsf-engines.

12. "About NSF Engines."

13. Dr. Giorgio Rizzoni, personal interview with author, July 27, 2023.

14. "Ohio State Awarded $3.8M Grant from U.S. Department of Energy to Improve Electric Vehicle Batteries," Ohio State News, January 20, 2023, https://news.osu.edu/ohio-state-awarded-38m-grant-from-us-department-of-energy-to-improve-electric-vehicle-batteries/.

15. "Overcoming EV Battery Manufacturing Challenges," Atlas Copco, accessed August 15, 2023, https://www.atlascopco.com/en-us/itba/expert-hub/articles/overcoming-ev-battery-manufacturing-challenges.

16. Robert Charette, "The EV Transition Explained: Battery Challenges," IEEE Spectrum, November 19, 2022, https://spectrum.ieee.org/the-ev-transition-explained-2658463682.

17. Joelle Anselmo, "Ohio State, Honda Partner on EV Battery R&D Center," *Manufacturing Dive*, November 29, 2023, https://www.manufacturingdive.com/news/osu-partners-with-honda-schaeffler-to-build-battery-research-center/700873/.

18. "LG Energy Solution and Honda Break Ground for New Joint Venture EV Battery Plant in Ohio," Honda, Electrification News, February 28, 2023, https://hondanews.com/en-US/electrification/releases/release-74895511bca6e7abc42504d7580c4aa4-lg-energy-solution-and-honda-break-ground-for-new-joint-venture-ev-battery-plant-in-ohio.

19. Yuri Kageyama, "Sony Apologizes for Battery Recall," *Washington Post*, October 24, 2006, https://www.washingtonpost.com/wpdyn/content/article/2006/10/24/AR2006102400278_pf.html.

Chapter 8

1. Leonard Fanning, *Carl Benz: Father of the Automobile Industry* (New York: Mercer Publishing, 1955).

2. Tom Nicholas and Vasiliki Fouka, "John D. Rockefeller: The Richest Man in the World," Harvard Business School Case 815–088, December 2014 (revised March 2018)

3. *The Derrick's Handbook of Petroleum: A Complete Chronological and Statistical Review of Petroleum Developments from 1859 to 1899* (Oil City, PA: Derrick Publishing Co., 1899), 1024.

4. Brita Åsbrink, "The War over the Oil Market," Nobel Brothers, August 15, 2011, https://www.branobelhistory.com/production/the-war-over-the-oil-market/.

5. "Papers of James Fortescue Flannery," Essex Record Office, Reference d/DU 861, accessed October 30, 2023, https://www.essexarchivesonline.co.uk/Result_Details.aspx?DocID=172588.

6. Peter B. Doran, *Breaking Rockefeller: The Incredible Story of the Ambitious Rivals who Toppled an Oil Empire* (New York: Penguin Books, 2016).

7. William Stevens, "Gas Stations: Tuning Up for the 1980s," *New York Times*, February 7, 1982, 4.

8. "1970s Energy Crisis," Wikipedia, accessed August 3, 2023, https://en.wikipedia.org/wiki/1970s_energy_crisis.

9. "Lithium Battery Basics: What's Inside a Lithium-Ion Battery?," DragonFly Energy, May 21, 2021, https://dragonflyenergy.com/inside-lithium-ion/.

10. Fred Lambert, "Tesla Model 3: Exclusive First Look at Tesla's New Battery Pack Architecture," *Electrek*, August 24, 2017, https://electrek.co/2017/08/24/tesla-model-3-exclusive-battery-pack-architecture/.

11. Lindsay Brooke, "Architecting GM's new EV batteries," SAE International, August 31, 2020, https://www.sae.org/news/2020/08/architecting-gms-new-ev-batteries.

12. Autumn Bows, "How the GM Ultium Compares to a Tesla Battery," *Hotcars*, July 24, 2020, https://www.hotcars.com/gm-ultium-compares-tesla-battery/.

13. "LG Chem," Wikipedia, https://en.wikipedia.org/wiki/LG_Chem (February 14, 2023).

14. Tom Jervis and Ricard Irving, "Tesla Model 3 Range, Battery and Charging," Driving Electric, October 30, 2023, https://www.drivingelectric.com/tesla/model-3/range.

15. "Alternative Fuels Data Center," United States Department of Energy, https://afdc.energy.gov/vehicles/electric_emissions.html (August 12, 2023).

16. "US Electricity Profile 2021," U.S. Energy Information Administration, November 2, 2023, https://www.eia.gov/electricity/state/.

17. "Wind Market Reports: 2022 Edition," The U.S. Department of Energy, https://www.energy.gov/eere/wind/wind-market-reports-2022-edition#wind (August 10, 2023).

18. Rosamond Hutt, "This Dutch Football Stadium Creates Its Own Energy and Stores It in Electric Car Batteries," World Economic Forum, July 6, 2018, https://www.weforum.org/agenda/2018/07/netherlands-football-johan-cruijff-stadium-electric-car-batteries/.

19. Dieter Holger and Giulia Petroni, "Old Electric-Vehicle Batteries Are Getting a Second Life," *Wall Street Journal*, June 13, 2022, https://www.wsj.com/articles/old-electric-vehicle-batteries-are-getting-a-second-life-11655114401.

20. Jason Thomas, "Work Begins on $3.5B EV Battery Manufacturing Plant in Berkeley County," *Charleston Regional Business Journal*, December 14, 2022.

21. Yasemin Esmen, "US Dives into Lithium Battery Recycling: 2023 Preview," Fastmarkets, January 4, 2023, https://www.fastmarkets.com/insights/us-dives-into-lithium-battery-recycling.

22. "AI for EV Battery Production and Recycling Optimization," Industry Today, March 17, 2023, https://industrytoday.com/ai-for-ev-battery-production-and-recycling-optimization.

23. Daniel de Visé, "Why Most Plastic Isn't Getting Recycled," *The Hill*, November 1, 2022, https://thehill.com/changing-america/sustainability/environment/3712999-why-most-plastic-isnt-getting-recycled/.

24. All data on hydrogen drawn from the International Energy Agency at https://www.iea.org.

25. Frank Markus, "1966 GM Electrovan Fuel Cell Prototype Turns 50," *Motor Trend*, November 1, 2016, https://www.motortrend.com/news/1966-gm-electrovan-fuel-cell-prototype-turns-50/.

26. Brayden Wood, "Why Hydrogen Electric Vehicles Will Be the Future of Auto Industry," *Top Speed*, June 26, 2023, https://www.topspeed.com/hydrogen-electric-vehicles-future-of-auto-industry/.

27. Hans Greimel, "Toyota Ponders Plug-In, Fuel Cell, Even an Electric Land Cruiser for the Future," *Automotive News*, August 11, 2023, https://www.autonews.com/cars-concepts/toyota-ponders-plug-fuel-cell-even-electric-land-cruiser.

28. Greimel, "Toyota Ponders Plug-In."

29. David Gelles, Brad Plumer, Jim Tankersley, and Jack Ewing, "The Clean Energy Future Is Arriving Faster Than You Think," *New York Times*, August 17, 2023, https://www.nytimes.com/interactive/2023/08/12/climate/clean-energy-us-fossil-fuels.html.

30. Steven Johnson, "The Brilliant Inventor Who Made Two of History's Biggest Mistakes," *New York Times*, March 17, 2023, https://www.nytimes.com/2023/03/15/magazine/cfcs-inventor.html.

Chapter 9

1. Leonard M. Fanning, *Carl Benz: Father of the Automobile Industry*. New York: Mercer Publishing, 1955.

2. "Carbon Black Market Size, Share & Growth Report [2030]." *Fortune Business Insights*, https://www.fortunebusinessinsights.com/industry-reports/carbon-black-market-101718.

3. Kasha Patel, "Why Tires—Not Tailpipes—Are Spewing More Pollution from Your Cars," *Washington Post*, July 9, 2023, https://www.washingtonpost.com/climate-environment/2023/07/09/tire-brake-tailpipes-emissions-pollution-cars/.

4. "Goodyear to Develop Domestic Source of Natural Rubber," Goodyear, April 7, 2022, https://corporate.goodyear.com/us/en/media/news/goodyear-to-develop-domestic-source-of-natural-rubber.html.

5. "Hallam Nuclear Power Facility." Wikipedia, https://en.wikipedia.org/wiki/Hallam_Nuclear_Power_Facility.

6. David Tracy, "Here's Why Tires Are Black," *Jalopnik*, November 15, 2018, https://jalopnik.com/heres-why-tires-are-black-1828925177.

7. Rob Hanson, personal interview with author, June 7, 2023. Note—all other direct quotes of Mr. Hanson are from the same interview.

8. Michael McCoy, "Monolith Plans Shake-Up of Carbon Black World," *Chemical & Engineering News*, May 18, 2015, https://cen.acs.org/articles/93/i20/Monolith-Plans-ShakeCarbon-Black-World.html.

9. McCoy, "Monolith Plans Shake-Up."

10. Christian Hinton, "Goodyear Unveils Carbon Black Tire Made with Renewable Energy," *Tire Review*, May 18, 2023, https://www.tirereview.com/goodyear-carbon-black-tire.

11. Sonal Patel, "NPPD to Convert Boiler to Use Hydrogen Fuel," *Power Magazine*, April 22, 2015, https://www.powermag.com/nppd-to-convert-coal-boiler-to-use-hydrogen-fuel/.

12. Aravind Chandrasekaran, Kevin Linderman, and Roger Schroeder, "Antecedents to Ambidexterity Competency in High Technology Organizations," *Journal of Operations Management* 30, nos. 1–2, (January 2012): 134–51.

13. Kelly Delaney-Klinger, Kenneth K. Boyer, Mark Frohlich, "The Return of Online Shopping: A Comparative Analysis of Webvan and Tesco's Operational Methods," *TQM Magazine*, June 1, 2003, 210–25.

14. Stefan Schneider, Sigfried Bajohr, Frank Graf, Thomas Kolb, "State of the Art of Hydrogen Production via Pyrolysis of Natural Gas," *Chemical & Biological Engineering Reviews* 7, no. 5, (July 13, 2020): 1–10.

15. Leah C. Stokes, "What Joe Manchin Cost Us," *New York Times*, July 16, 2022, https://www.nytimes.com/2022/07/16/opinion/environment/joe-manchin-climate-change-clean-energy.html.

16. "Goodyear Collaborates with Monolith on Carbon Black; Initial Testing Demonstrates Reduced Emissions," Goodyear, December 9, 2021, https://corporate.goodyear.com/us/en.https://corporate.goodyear.com/us/en/media/news/goodyear-collaborates-with-monolith-on-carbon-black-initial-testing-demonstrates-reduced-emissions.html.

17. Rob Hanson, personal interview with author, October 31, 2023.

18. "Advanced Drainage Systems Releases Fiscal Year 2022 Sustainability Report," ADS, August 21, 2022, https://www.adspipe.com/the-ripple-effect/sustainability/advanced-drainage-systems-releases-fiscal-year-2022-sustainability-report.

19. Rob Hanson, personal interview with author, July 8, 2023.

20. Alyssa Young, personal interview with author, July 21, 2023.

21. Megan Smalley, "Liberty Tire Recycling, Bolder Industries Expand Partnership," *Recycling Today*, August 2, 2022, https://www.recyclingtoday.com/news/liberty-tire-recycling-bolder-formalize-partnership/.

Chapter 10

1 . John Tobin, "Henry Ford and His Village Industries in Southeastern Michigan," (unpublished thesis, Eastern Michigan University, 1985), 196–211.

2. Mitchell Bilo, "Ford's Village Industries," *Pittsburgh Press*, November 8, 1988, 34.

3. "Village industries," Wikipedia, accessed August 13, 2023, https://en.wikipedia.org/wiki/Village_industries.

4. "Norway—the EV Capital of the World," Visit Norway, accessed August 15, 2023, https://www.visitnorway.com/plan-your-trip/getting-around/by-car/electric-cars/.

5. Anthony DeSimone and John J. Sullivan IV, "Where the Water Meets the Road," *Public Roads* 77, no. 5 (March–April 2014), https://highways.dot.gov/public-roads/marchapril-2014/where-water-meets-road; "Transport in Norway: Ferries," Wikipedia, https://en.wikipedia.org/wiki/Transport_in_Norway#Ferries; James Ayre "Feasible to Replace 70% of Norwegian Ferries with Fully Electric or Hybrid Ferries," CleanTechnica, July 27, 2016.

6. "Countries in the World by Population," World o Meter, July 16, 2023, https://www.worldometers.info/world-population/population-by-country/.

7. "EV Charging Infrastructure Best Practice: What We Can Learn from Norway," Mer UK, February 1, 2024, https://uk.mer.eco/news/ev-charging-infrastructure-best

-practice-learnings-from-norway/; "Discover Norway's Unique EV and EV Chargers Perks," Wallbox, February 1, 2024, https://blog.wallbox.com/norway-ev-incentives/.

8. "Electric Vehicle Chargers in California," California Energy Commission, accessed August 15, 2023, https://www.energy.ca.gov/data-reports/energy-almanac/zero-emission-vehicle-and-infrastructure-statistics/electric-vehicle.

9. Jack Ewing, "In Norway, the Electric Vehicle Future Has Already Arrived," *New York Times*, May 8, 2023, https://www.nytimes.com/2023/05/08/business/energy-environment/norway-electric-vehicles.html.

10. Scooter Doll, "Volvo Braces for Challenging Year Ahead after More Than Doubling EV Sales and Breaking Revenue Record in 2022," *Electrek*, February 9, 2023, https://electrek.co/2023/02/09/volvo-challenging-year-doubling-ev-sales-breaking-revenue-record-2022.

11. Ewing, "In Norway, the Electric Vehicle Future Has Already Arrived."

12. Ewing.

13. Ewing.

14. Thomas L. Friedman, *Hot, Flat, and Crowded: Why We Need a Green Revolution—and How It Can Renew America* (New York: Farrar, Straus and Giroux, 2008)

15. Nick Gibbs, "Sony Honda JV plans decade-long lease for high-tech EV," *Automotive News*, January 13, 2023, https://www.autonews.com/automakers-suppliers/sony-honda-jv-plans-decade-long-lease-high-tech-ev.

16. Zachary Abuza, "Vietnam's VinFast Appears to Be Going Places, but Its US Entry Reveals Issues under the Hood," *South China Morning Post*, August 31, 2023, https://www.scmp.com/week-asia/opinion/article/3232936/vietnams-vinfast-appears-be-going-places-its-us-entry-reveals-issues-under-hood#.

Bibliography

"About NSF Engines—Regional Innovation Engines." | National Science Foundation." https://new.nsf.gov/funding/initiatives/regional-innovation-engines/about-nsf-engines.

Abuza, Zachary. "Vietnam's VinFast Appears to Be Going Places, but Its US Entry Reveals Issues under the Hood." *South China Morning Post*, August 31, 2023. https://www.scmp.com/week-asia/opinion/article/3232936/vietnams-vinfast-appears-be-going-places-its-us-entry-reveals-issues-under-hood#

"Advanced Drainage Systems Releases Fiscal Year 2022 Sustainability Report." ADS, August 21, 2022. https://www.adspipe.com/the-ripple-effect/sustainability/advanced-drainage-systems-releases-fiscal-year-2022-sustainability-report.

"AI for EV Battery Production and Recycling Optimization." *Industry Today*, March 17, 2023. https://industrytoday.com/ai-for-ev-battery-production-and-recycling-optimization/.

AlMulla, Yasmena. "Kuwait: Fire Ripped through 25,000 Square Meters of Al Sulabiya Tire Site." *Gulf News*, October 23, 2020. https://gulfnews.com/world/gulf/kuwait/kuwait-fire-ripped-through-25000-square-meters-of-al-sulabiya-tire-site-1.74769506.

Amend, James M. "Mary Barra Appointment Puts Insider Back in GM CEO's Seat." *WardsAuto*, December 1, 2013. https://www.wardsauto.com/industry/mary-barra-appointment-puts-insider-back-gm-ceo-s-seat.

Arlidge, John. "Mary the Mechanic Pulls GM out of the Wreckage," *Sunday Times* (London), September 20, 2015.

Åsbrink, Brita. "The War over the Oil Market." Nobel Brothers, August 15, 2011. https://www.branobelhistory.com/production/the-war-over-the-oil-market/.

"Auto Strike Settlements Will Raise Costs for Detroit's Big 3." *Business Inquirer*, November 1, 2023. https://business.inquirer.net/429598/auto-strike-settlements-will-raise-costs-for-detroits-big-3.

Ayre, James. "Feasible to Replace 70% of Norwegian Ferries with Fully Electric or Hybrid Ferries." CleanTechnica, July 27, 2016. https://cleantechnica.com/2016/07/27/feasible-replace-70-norwegian-ferries-battery-hybrid-ferries-study/.

Bailey, Kat. "The DeLorean Is Officially Back, and It's Electric." IGN Southeast Asia. February 15, 2022. https://sea.ign.com/back-to-the-future-theater/181994/news/the-delorean-is-officially-back-and-its-electric.

Beltran, Luisa. "VinFast, the Latest EV IPO, Comes to U.S. from Vietnam." *Barrons*, April 7, 2022. https://www.barrons.com/articles/vinfast-ev-ipo-vietnam-51649367803.

Benchmark Mineral Intelligence. *Ohio Battery Supply Chain Opportunities*, JobsOhio, February 2022, https://cdn.bfldr.com/AHJE351Z/at/cx67vf3tm69g4r49w5v3z/BMI_Ohio_Report.pdf.

Bilo, Mitchell. "Ford's Village Industries." *Pittsburgh Press*, November 8, 1988, 34.

"Black Spade Acquisition Co Announces Extension of Combination Period." *PR Newswire*, July 14, 2023. https://www.prnewswire.com/news-releases/black-spade-acquisition-co-announces-extension-of-combination-period-301877405.html.

Boudette, Neal E. 2018. "For Tesla, 'Production Hell' Looks Like the Reality of the Car Business." *New York Times*, April 3, 2018. https://www.nytimes.com/2018/04/03/business/tesla-model-3.html.

Bows, Autumn. "How the GM Ultium Compares to a Tesla Battery." *HotCars*, July 24, 2020. https://www.hotcars.com/gm-ultium-compares-tesla-battery/.

Brinkley, Douglas. 2023. "Prime Mover." *American Heritage* 54 (3). https://www.americanheritage.com/prime-mover.

Brooke, Lindsay. "Architecting GMs New EV Batteries." SAE International, August 31, 2020. https://www.sae.org/site/news/2020/08/architecting-gms-new-ev-batteries.

Brynjolfsson, Erik, and Michael D. Smith. "Frictionless Commerce? A Comparison of Internet and Conventional Retailers." *Management Science* 46, no. 4 (2000): 563–85. doi:10.1287/mnsc.46.4.563.12061. http://mansci.journal.informs.org/cgi/content/abstract/46/4/563.

Bushey, Claire. "GM Tries to Catch Tesla by Following Its Supply Chain Playbook." *Financial Times*, June 18, 2023. https://www.ft.com/content/8e9b17d3-5e1c-4aad-9e28-e6b96648dd7b.

Caldwell, A. J. "Why Top Automakers Spend Millions on Concept Cars they Don't Plan on Making." *Business Insider*, January 27, 2021. https://www.businessinsider.com/automakers-spend-millions-on-concept-cars-they-dont-make-2019-4.

Campbell, Jaelyn. "VinFast and Yorkville Reach $1 Billion Share Deal." CBT News, October 20, 2023. https://www.cbtnews.com/vinfast-and-yorkville-reach-1-billion-share-deal/.

"Carbon Black Market Size, Share & Growth Report 2023–2030." *Fortune Business Insights*, https://www.fortunebusinessinsights.com/industry-reports/carbon-black-market-101718.

"Center for Automotive Research." Center for Automotive Research, July 28, 2023. https://car.osu.edu/.

Chandrasekaran, Aravind, Kevin Linderman, and Roger Schroeder. "Antecedents to Ambidexterity Competency in High Technology Organizations." *Journal of Operations Management* 30, nos. 1–2 (2012): 134–51. doi:10.1016/j.jom.2011.10.002. https://dx.doi.org/10.1016/j.jom.2011.10.002.

Charette, Robert N. "The EV Transition Explained: Battery Challenges." IEEE Spectrum, November 19, 2022. https://spectrum.ieee.org/the-ev-transition-explained-2658463682.

Colias, Mike. "GM to Recover $1.9 Billion in Bolt-Recall Costs in Deal with LG." *Wall Street Journal*, October 12, 2021. https://www.wsj.com/articles/gm-says-it-will-recover-1-9-billion-of-bolt-recall-costs-from-lg-11634042381.

"Countries in the World by Population." World o Meter, July 16, 2023. https://www.worldometers.info/world-population/population-by-country/.

Cox, Christopher. "Elon's Appetite for Destruction." *New York Times*, January 17, 2023. https://www.nytimes.com/2023/01/17/magazine/tesla-autopilot-self-driving-elon-musk.html.

"'Crown Jewel': Electric Carmaker Vinfast Breaks Ground in Its $4B NC Manufacturing Plant." *ABC 11 News*. July 28, 2023. https://abc11.com/vinfast-north-carolina-chatham-county-roy-cooper-electric-vehicle/13562386/.

Cyran, Robert. "The iPad Factor in Digital Readers." *New York Times*, June 28, 2010. https://www.nytimes.com/2010/06/28/technology/28views.html.

Daga, Anshuman, and Phuong Nguyen. Vietnam's VinFast Taps Banks for $4 Bln EV Factory Funding Deal." Reuters, July 13, 2022. https://www.reuters.com/business/autos-transportation/vietnams-vinfast-says-agrees-4-bln-ev-factory-funding-with-credit-suisse-citi-2022-07-13/.

Delaney-Klinger, Kelly, Kenneth K. Boyer, and Mark Frohlich. "The Return of Online Grocery Shopping: A Comparative Analysis of Webvan and Tesco's Operational Methods." *TQM Magazine* 15, no. 3 (2003): 210–55. doi:10.1108/09544780310469334. https://www.proquest.com/scholarly-journals/return-online-grocery-shopping-comparative/docview/227576212/se-2?accountid=9783.

The Derrick's Handbook of Petroleum: A Complete Chronological and Statistical Review of Petroleum Developments from 1859 to 1899 1899. Oil City, PA: Derrick Publishing Co.

DeSimone, Anthony, and John J. Sullivan IV. "Where the Water Meets the Road." *Public Roads* 77, no. 5 (March–April 2014). https://highways.dot.gov/public-roads/marchapril-2014/where-water-meets-road.

Deters, Henning. "European Environmental Policy at 50: Five Decades of Escaping Decision Traps?" *Environmental Policy and Governance* 29 (May 17, 2019): 315–25. doi:10.1002/EET.1855.

de Visé, Daniel. "Why Most Plastic Isn't Getting Recycled." *The Hill*, November 1, 2022. https://thehill.com/changing-america/sustainability/environment/3712999-why-most-plastic-isnt-getting-recycled/.

Doll, Scooter. "Ford F-150 Lightning Gets $10K Price Cut, $50K MSRP as Ramping Supply Meets Demand." *Electrek*, July 17, 2023. https://electrek.co/2023/07/17/ford-f-150-lightning-production-scale-price-cut-near-10000-new-customers/.

Donut, "We Drove the Worst Reviewed Car in America." YouTube, July 12, 2023. https://www.youtube.com/watch?v=DF7kaLTsNHQ.

Doran, Peter B. *Breaking Rockefeller: The Incredible Story of the Ambitious Rivals Who Toppled an Oil Empire*. New York: Penguin Books, 2016.

Dragonfly Energy. "Lithium Battery Basics: What's Inside A Lithium-Ion Battery?" May 21, 2021. https://dragonflyenergy.com/inside-lithium-ion/.

"Editorial: EV Battery Boom: Manchin Gambit Might be Working." *Automotive News*, October 16, 2022. https://www.autonews.com/editorial/ev-battery-boom-joe-manchins-gambit-might-be-working.
Editors, K. B. B. "Honda Hybrid Sales Reach 1 Million Worldwide." *Kelley Blue Book*, October 16, 2012. https://www.kbb.com/car-news/honda-hybrid-sales-reach-1-million-worldwide/.
Edmonds, Ellen. "Nearly 51 Million Americans to Travel This Thanksgiving, Highest Volume in a Dozen Years." AAA Newsroom, November 16, 2017. https://newsroom.aaa.com/2017/11/nearly-51-million-americans-travel-thanksgiving-highest-volume-dozen-years/.
"Edsel," Wikipedia. Accessed 16, 2023. https://en.wikipedia.org/wiki/Edsel.
"Electric Vehicle Chargers in California," California Energy Commission. Accessed August 15, 2023. https://www.energy.ca.gov/data-reports/energy-almanac/zero-emission-vehicle-and-infrastructure-statistics/electric-vehicle.
Esmen, Yasemin. "US Dives into Lithium Battery Recycling: 2023 Preview." Fastmarkets, January 4, 2023. http://fastmarkets-prod-01.altis.cloud/insights/us-dives-into-lithium-battery-recycling/.
Ewing, Jack. "Electric Car Sales Climb Sharply Despite Shortages." *New York Times*, July 14, 2022. https://www.nytimes.com/2022/07/14/business/electric-car-sales.html?referringSource=articleShare.
———. "In Norway, the Electric Vehicle Future Has Already Arrived." *New York Times*, May 8, 2023, https://www.nytimes.com/2023/05/08/business/energy-environment/norway-electric-vehicles.html.
———. "Tesla May Already Have Won the Charging Wars." *New York Times*, June 27, 2023. https://www.nytimes.com/2023/06/27/business/energy-environment/tesla-gm-ford-charging-electric-vehicles.html.
Fanning, Leonard. *Carl Benz: Father of the Automobile Industry*. New York: Mercer Publishing, 1955.
Ferris, Robert. "Tesla Will Need Billions to Make US Supercharger Network Compete with Gas Stations, Says Analyst." CNBC, March 3, 2017. https://www.cnbc.com/2017/03/03/tesla-needs-billions-to-make-supercharger-network-rival-gas-stations.html.
Ford Motor Company. *1909–1927 Ford Model T, TT Service Manual*. Detroit: Detroit Iron, 1927.
Friedman, Alexi. "Former GM Plant Is Razed for Linden 'Renaissance.'" *Star-Ledger* (New Jersey), October 31, 2023. https://www.nj.com/news/2008/08/former_gm_plant_is_razed_for_l.html.
Fung, Brian. "The Chevy Bolt Aims to Dethrone Tesla. Here Are Our First Impressions." *Washington Post*, September 21, 2016. https://www.washingtonpost.com/news/the-switch/wp/2016/09/21/the-chevy-bolt-aims-to-dethrone-tesla-here-are-our-first-impressions/.
Gates, Andi. "Supermarket Sells Eggs and Equities." *Columbus Dispatch*, March 27, 1983.

Gelles, David, Brad Plumer, Jim Tankersley, and Jack Ewing. "The Clean Energy Future Is Arriving Faster Than You Think." *New York Times*, August 17, 2023. https://www.nytimes.com/interactive/2023/08/12/climate/clean-energy-us-fossil-fuels.html.

"General Motors Takes Q3 Sales Crown, but GM Stock Makes New Low." *Investor's Business Daily*, October 4, 2023. https://www.investors.com/news/auto-sales-q3-2023-gm-ford-gains/.

Ghosal, Anuruddha. "Vietnam's VinFast Committed to Selling EVs to US Despite Challenges, Intense Competition." ABC News, October 24, 2023. https://abcnews.go.com/International/wireStory/vietnams-vinfast-committed-selling-evs-us-despite-challenges-104269388.

Giang, Nguyen Kieu. "Vietnamese EV Maker VinFast Files for US IPO." *Bloomberg*, December 6, 2022. https://www.bloomberg.com/news/articles/2022-12-06/vietnamese-electric-vehicle-maker-vinfast-files-for-us-ipo.

"GM Builds Electric Battery Lab in Michigan as It Tries to Cut EV Costs, Extend." *Washington Post*, October 5, 2021. https://www.washingtonpost.com/business/economy/gm-builds-electric-battery-lab-in-michigan-as-it-tries-to-cut-ev-costs-extend-range/2021/10/05/320d5186-25cb-11ec-8831-a31e7b3de188_story.html.

"GM CEO Mary Barra Talks Sustainability with Chelsea Clinton during Climate Week." 3BL, September 26, 2014. https://www.3blmedia.com/news/gm-ceo-mary-barra-talks-sustainability-chelsea-clinton-during-climate-week.

"GM Reports Record Net Income of $9.7 Billion and Record EBIT-Adjusted of $10.8 Billion for 2015." General Motors Newsroom. Accessed August 7, 2023. https://news.gm.com/public/us/en/gm/home/newsroom.detail.html/Pages/news/emergency_news/2016/0203-2015-4th-qtr-earnings.html.

"Goodyear Collaborates with Monolith on Carbon Black; Initial Testing Demonstrates Reduced Emissions." Goodyear, December 9, 2021. https://corporate.goodyear.com/us/en.https://corporate.goodyear.com/us/en/media/news/goodyear-collaborates-with-monolith-on-carbon-black-initial-testing-demonstrates-reduced-emissions.html.

"Goodyear to Develop Domestic Source of Natural Rubber." Goodyear, April 7, 2022. https://corporate.goodyear.com/us/en.

Gordon, Brian, and Tyler Dukes. "VinFast Lobbied Gov. Cooper to Advance Federal Loan Bid to Avoid 'Further' NC Delays." *News & Observer*, August 8, 2023. https://www.newsobserver.com/news/politics-government/article277702563.html.

"Governor DeWine Announces Honda to Invest in Ohio for Electric Vehicle Production, Including New Battery Plant with LG Energy Solution." Off ice of the Governor of Ohio. Press release, October 11, 2022. https://governor.ohio.gov/media/news-and-media/governor-dewine-announces-honda-to-invest-in-ohio-for-electric-vehicle-production-including-new-battery-plant-with-lg-energy-solution-10112022.

"Greenhouse Gas Emissions from a Typical Passenger Vehicle." Environmental Protection Agency, August 28, 2023. https://www.epa.gov/greenvehicles/greenhouse-gas-emissions-typical-passenger-vehicle.

Greimel, Hans. "Honda CEO Toshihiro Mibe Outlines Multistep Plan to Jump Ahead in EVs." *Automotive News*, April 26, 2023. https://www.autonews.com/automakers-suppliers/honda-ceo-pledges-fight-back-behind-ev-race.

———. "Toyota Ponders Plug-In, Fuel Cell, Even an Electric Land Cruiser for the Future." *Automotive News*, August 11, 2023. https://www.autonews.com/cars-concepts/toyota-ponders-plug-fuel-cell-even-electric-land-cruiser.

Hawkins, Andrew J. "GM and Honda Expanding their Partnership to Develop 'Millions' of Affordable EVs in 2027." *Verge*, April 5, 2022. https://www.theverge.com/2022/4/5/23011201/gm-honda-ev-partnership-millions-affordable-2027.

———. "GM Unveils New Logo to Emphasize its Pivot to Electric Vehicles." *Verge*, January 8, 2021. https://www.theverge.com/2021/1/8/22220574/gm-new-logo-electric-vehicle-brand-ces-2021.

Hinton, Christian. "Goodyear Unveils Carbon Black Tire Made with Renewable Energy." *Tire Review*, May 18, 2023. https://www.tirereview.com/goodyear-carbon-black-tire/.

Holger, Dieter, and Giulia Petroni. "Old Electric-Vehicle Batteries Are Getting a Second Life." *Wall Street Journal*, June 13, 2022. https://www.wsj.com/articles/old-electric-vehicle-batteries-are-getting-a-second-life-11655114401.

Holt, Kris. "GM Modernizes Its Logo to Highlight Its EV-Centric Future." Engadget, January 8, 2021. https://www.engadget.com/gm-logo-branding-electric-vehicles-210607845.html.

Holweg, Matthias. "The Genealogy of Lean Production." *Journal of Operations Management* 25, no. 2 (2006): 420–37. https://doi.org/10.1016/j.jom.2006.04.001.

"Honda." Wikipedia. Accessed July 11, 2023. https://en.wikipedia.org/wiki/Honda.

"Honda Makes Major Investment in Ohio to Create New Electric Vehicle Hub." Honda Newsroom, October 11, 2022. http://hondanews.com/en-US/releases/honda-makes-major-investment-in-ohio-to-create-new-electric-vehicle-hub.

"Honda's Marysville Auto Plant at 25 Years." Honda, August 21, 2007. http://hondanews.com/en-US/releases/hondas-marysville-auto-plant-at-25-years-historic-yet-new.

"How GM Destroyed Its Saturn Success." *Forbes*, March 8, 2010. https://www.forbes.com/2010/03/08/saturn-gm-innovation-leadership-managing-failure.html.

Hughes, Siobhan. "Senators Challenge GM's Barra, Push for Faster Change." *Wall Street Journal*, April 2, 2014. https://www.wsj.com/articles/gm-faces-new-round-of-tough-questions-on-capitol-hill-1405609879.

Hutt, Rosamond. "This Dutch Football Stadium Creates Its Own Energy and Stores It in Electric Car Batteries," World Economic Forum, July 6, 2018. https://www.weforum.org/agenda/2018/07/netherlands-football-johan-cruijff-stadium-electric-car-batteries/.

Hyde, Ellis. "Do Electric Cars Have Gearboxes?" DrivingElectric, last modified January 30, https://www.drivingelectric.com/your-questions-answered/95/do-electric-cars-have-gearboxes.

Iliff, Laurence. "VinFast Considers Ultralow-Priced VF 3 EV for U.S. at Urging of Prospective Dealers." *Automotive News*, October 27, 2023. https://www.autonews.com/retail/vinfast-gauges-dealer-interest-sub-20000-ev.

"Inflation Reduction Act of 2022." US Department of Energy, July 14, 2023. https://www.energy.gov/lpo/inflation-reduction-act-2022.

Irwin, John, and Kenney, David. "Magna Pegs $700M for EV Battery Enclosures." *Automotive News*, February 18, 2023. https://www.autonews.com/suppliers/magna-spend-michigan-ontario-bid-be-big-ev-player.

"Jealousy Quotes." Goodreads. https://www.goodreads.com/quotes/tag/jealousy.

Jervis, Tom, and Richard Ingram. "Tesla Model 3: Range, Battery and Charging 2023." Driving Electric, October 30, 2023. https://www.drivingelectric.com/tesla/model-3/range.

"John DeLorean." Wikipedia. Accessed July 14, 2023. https://en.wikipedia.org/wiki/John_DeLorean.

Johnson, Steven. "The Brilliant Inventor Who Made Two of History's Biggest Mistakes." *New York Times Magazine*, March 17, 2023. https://www.nytimes.com/2023/03/15/magazine/cfcs-inventor.html.

Kageyama, Yuri. "Sony Apologizes for Battery Recall," *Washington Post*, October 24, 2006, https://www.washingtonpost.com/wpdyn/content/article/2006/10/24/AR2006102400278_pf.html.

"Kaiser-Frazer." Wikipedia. Accessed July 16, 2023. https://en.wikipedia.org/wiki/Kaiser-Frazer.

Kane, Mark. "US: GMC Hummer EV Pickup Sales Down 83 Percent in Q2 2023." *InsideEVs*, July 05, 2023. https://insideevs.com/news/675394/us-gmc-hummer-ev-sales-2023q2/.

Katakis, Manoli. "Executive Vice President of General Motors Global Manufacturing Tim Lee to Retire April 1." GM Authority, January 26, 2014. https://gmauthority.com/blog/2014/01/executive-vice-president-of-general-motors-global-manufacturing-tim-lee-to-retire-april-1/.

Keenan, Tim. "American Battery Solutions Acquires Tiveni's Entire Battery Systems IP Portfolio." *DBusiness*, September 14, 2022. https://www.dbusiness.com/daily-news/american-battery-solutions-acquires-tivenis-entire-battery-systems-ip-portfolio/.

Klyayman, Ben. "GM Aims to Build 500,000 Electric-Technology Vehicles a Year." Reuters, November 14, 2012. https://www.reuters.com/article/us-gm-evs/gm-aims-to-build-500000-electric-technology-vehicles-a-year-idUKBRE8AD1EM20121114/.

Korn, Morgan. "Hummer Electric Pickup Truck Unveiled: What You Need to Know." ABC News, October 20, 2020. https://abcnews.go.com/Business/hummer-electric-pickup-truck-unveiled/story?id=73693774.

Lambert, Fred. "Tesla Model 3: Exclusive First Look at Tesla's New Battery Pack Architecture." *Electrek*, August 24, 2017. https://electrek.co/2017/08/24/tesla-model-3-exclusive-battery-pack-architecture/.

———. "Tesla Tries to Get Owners to Give Up 'Unlimited Free Supercharging for Life.'" *Electrek*, April 24, 2023. https://electrek.co/2023/04/24/tesla-triesowners-give-up-unlimited-free-supercharging-for-life/.

Lambrecht, Andrew. "VinFast and Black Spade Announce $27B Merger for August IPO." *InsideEVs*, July 31, 2023. https://insideevs.com/news/679514/vinfast-secured-by-sec/.

LaMonica, Martin. "Tesla Motors Founders: Now There Are Five." CNET, September 21, 2009. https://www.cnet.com/culture/tesla-motors-founders-now-there-are-five/.

LaReau, Jamie. "GM Forms Alliance with Honda to Develop Future Products in North America." *Detroit Free Press,* September 3, 2020. https://www.freep.com/story/money/cars/general-motors/2020/09/03/general-motors-honda-partnership-north-america/5701104002/#.

Laseter, Tim, Patrick W. Houston, Joshua L. Wright, and Juliana Y. Park. "Amazon Your Industry: Extracting Value from the Value Chain." *Strategy & Business,* January 1, 2000. https://www.strategy-business.com/article/10479.

Lasseter, Evan. "Average Prices for New EVs Are Falling: 'Still a Luxury' for Most, but That's Changing as More Are Produced," *Atlanta Journal-Constitution,* June 18, 2023.

Lenschow, Andrea, and Carina Sprungk. "The Myth of a Green Europe." *JCMS: Journal of Common Market Studies* 48, no. 1 (2009): 133–54. doi:10.1111/j.1468-5965.2009.02045.x. https://doi.org/10.1111/j.1468-5965.2009.02045.x.

Levin, Tim. "These Are the 20 Car Brands with the Most Loyal Customers." *Business Insider,* July 17, 2020. https://www.businessinsider.com/car-buying-brands-most-loyal-customers-automotive-sales-loyalty-subaru-2020-7.

Lewy, Guenter. *America in Vietnam.* New York: Oxford University Press, 1978.

"LG Chem." Wikipedia. Accessed February 14, 2023. https://en.wikipedia.org/wiki/LG_Chem.

"LG Energy Solution and Honda Break Ground for New Joint Venture EV Battery Plant in Ohio." Honda, Electrification News, February 28, 2023. http://hondanews.com/en-US/electrification/releases/lg-energy-solution-and-honda-break-ground-for-new-joint-venture-ev-battery-plant-in-ohio.

Lienert, Paul. "New GM CEO Barra Seen as a Winner Early in Her Career." Reuters, December 10, 2013. https://www.reuters.com/article/uk-autos-barra-personal/new-gm-ceo-barra-seen-as-a-winner-early-in-her-career-idUKBRE9BA00H20131211/.

"Little's Law." Wikipedia, December 18, 2023. https://en.wikipedia.org/wiki/Little%27s_law.

Loukas. "EV vs ICE—Differences and Similarities." ArenaEV, April 24, https://www.arenaev.com/ice_v_ev__differences_and_similarities-news-185.php.

Lutz, Hannah. "EV Transition Slows as Inventory Grows and Industry Hits Hurdles." Automotive News, October 31, 2023. https://www.autonews.com/retail/ev-transition-slows-amid-inventory-woes-sluggish-demand.

Maister, David. "The Psychology of Waiting Lines." *David Maister,* 1985. https://davidmaister.com/articles/the-psychology-of-waiting-lines/.

Markus, Frank. "1966 GM Electrovan Fuel Cell Prototype Turns 50." *Motor Trend,* November 1, 2016. https://www.motortrend.com/news/1966-gm-electrovan-fuel-cell-prototype-turns-50/.

Mast, Garrett. "Toyota to Collaborate with Redwood Materials on a Sustainable, Closed-Loop Electrified Vehicle Battery Ecosystem." Toyota USA Newsroom,

June 21, 2022. https://pressroom.toyota.com/toyota-to-collaborate-with-redwood-materials-on-a-sustainable-closed-loop-electrified-vehicle-battery-ecosystem/.

Matousek, Mark. "Mary Barra was Called a 'Lightweight' when She Became CEO of GM—Here's How She Transformed the Company and Silenced Her Doubters." *Business Insider*, January 11, 2018. https://www.businessinsider.com/heres-how-mary-barra-silenced-critics-who-called-her-a-lightweight-2018-1.

McCoy, Michael. "Monolith Plans Shake-Up of Carbon Black World." *Chemical & Engineering News*, May 18, 2015. https://cen.acs.org/articles/93/i20/Monolith-Plans-ShakeCarbon-Black-World.html.

McDonald, Joe. "China Auto Show Highlights Intense Electric Car Competition." Associated Press, April 15, 2023. https://apnews.com/article/china-auto-electric-vehicle-0ea339b7be33090b222169fd433cf51c.

McDonald, Loren. "UBS Analyst Gets Future Investment Costs for Tesla Supercharger Network Super Wrong." CleanTechnica, March 5, 2017. https://cleantechnica.com/2017/03/05/ubs-analyst-gets-future-investment-costs-tesla-supercharger-network-super-wrong/.

Meyer, Robinson. "America Can't Build a Green Economy without China." *New York Times*, July 17, 21023. https://www.nytimes.com/2023/07/17/opinion/america-china-clean-energy.html.

Mihalascu, Dan. "High Electrification Costs Are 'Elephant in the Room': Stellantis." *Inside EVs*, April 10, 2023. https://www.msn.com/en-us/autos/news/high-electrification-costs-are-elephant-in-the-room-stellantis/ar-AA19GrIv.

MikeL01. "VinFast to Stop Making Gas-Powered Cars by End-August." VinFast Forum, July 16, 2022. https://www.vinfasttalk.com/threads/vinfast-to-stop-making-gas-powered-cars-by-end-august.127/.

Mintzberg, Henry, Richard T. Pascale, Michael Goold, and Richard Rumelt. "The Honda Effect Revisited." *California Management Review* 34, no. 1 (Spring 1996): 78–91.

"Mobility at the Ohio State University." The Ohio State University. Accessed September 11, 2023. https://www.osu.edu/research/mobility/.

Nauss, Donald W. "GM's Man Who Bested NBC Helps Rouse Sleeping Giant: Business: Counsel Harry Pearce Wields Vast Power. He Is Seen as Best of New Talent Shaking Up Troubled Firm." *Los Angeles Times*, February 17, 1993. https://www.latimes.com/archives/la-xpm-1993-02-17-mn-238-story.html.

Nguyen, Phuong. "Vietnam's VinFast Recalls First Batch of US-Bound EVs Over Safety Risk." Reuters, May 5, 2023. https://www.reuters.com/business/autos-transportation/vietnams-vinfast-recalls-all-first-batch-evs-shipped-us-over-security-risk-2023-05-25/.

———. "VinFast Posts Deeper Loss in Q1, Eyes July Completion of SPAC Merger." Reuters, June 16, 2023. https://www.reuters.com/technology/vinfast-posts-deeper-loss-q1-eyes-july-completion-spac-merger-2023-06-16/.

———. "Vietnam's VinFast Ships First Electric Vehicles to U.S. Customers." Reuters, November 25, 2022. https://www.reuters.com/business/autos-transportation/vietnams-vinfast-ships-first-electric-vehicles-us-customers-2022-11-25/.

Nhan, Thanh. "Auto Newbie VinFast Zooms to Fifth Place in Sales." *VnExpress International* April 26, 2020. https://e.vnexpress.net/news/business/companies/auto-newbie-vinfast-zooms-to-fifth-place-in-sales-4090698.html.

Nicholas, Tom, and Vasiliki Fouka. "John D. Rockefeller: The Richest Man in the World." Harvard Business School, Case 815–088. https://www.hbs.edu/faculty/Pages/item.aspx?num=47167.

"1970s Energy Crisis." Accessed August 3, 2023. Wikipedia. https://en.wikipedia.org/wiki/1970s_energy_crisis.

Nitta, Yuji. "Vietnam Vingroup's Ambition to Take on Tesla Hits Bumpy Road." *Nikkei Asia*, August 3, 2023. https://asia.nikkei.com/Business/Automobiles/Vietnam-Vingroup-s-ambition-to-take-on-Tesla-hits-bumpy-road.

"Norway—the EV Capital of the World." Visit Norway. Accessed August 15, 2023. https://www.visitnorway.com/plan-your-trip/getting-around/by-car/electric-cars/.

"Ohio Battery Supply Chain Opportunities." 2022g.

"Ohio State Awarded $3.8M Grant from U.S. Department of Energy to Improve Electric Vehicle Batteries." Ohio Stat News, January 20, 2023. https://news.osu.edu/ohio-state-awarded-38m-grant-from-us-department-of-energy-to-improve-electric-vehicle-batteries/.

Ohnsman, Alan. "California's Lithium Rush for EV Batteries Hinges on Taming Toxic, Volcanic Brine." *Forbes*, last modified August 31, 2022. https://www.forbes.com/sites/alanohnsman/2022/08/31/californias-lithium-rush-electric-vehicles-salton-sea/.

O'Kane, Sean. "Lucid Motors Goes Public, Collects $4.5 Billion." *Verge*, July 26, 2021. https://www.theverge.com/2021/7/26/22594177/lucid-motors-spac-nasdaq-saudi-arabia-ev-startup.

Osaka, Shannon. "Elon Musk Agrees to Open Tesla's Charging Network to Non-Tesla Drivers." *Washington Post*, February 15, 2023. https://www.washingtonpost.com/climate-environment/2023/02/15/tesla-supercharger-network-locked/#.

"Overcoming EV Battery Manufacturing Challenges." Atlas Copco. Accessed August 15, 2023. https://www.atlascopco.com/en-us/itba/expert-hub/articles/overcoming-ev-battery-manufacturing-challenges.

"Papers of Sir James Fortescue Flannery M.P." Essex Record Office, 1917. https://www.essexarchivesonline.co.uk/Result_Details.aspx?DocID=172588.

Partsch, Elizabeth. Europe's Transition to Electric Vehicles: How It's Going, and What Lies Ahead. Impakter, December, 24, 2022. https://impakter.com/europes-transition-to-electric-vehicles-hows-it-going/.

Patel, Kasha. "Why Tires—Not Tailpipes—Are Spewing More Pollution from Your Cars." *Washington Post*, July 9, 2023. https://www.washingtonpost.com/climate-environment/2023/07/09/tire-brake-tailpipes-emissions-pollution-cars/.

Patel, Sonal. "NPPD to Convert Coal Boiler to Use Hydrogen Fuel." *Power Magazine*, April 22, 2015. https://www.powermag.com/nppd-to-convert-coal-boiler-to-use-hydrogen-fuel/.

Plumer, Brad, and Lisa Friedman. "A Swaggering Clean-Energy Pioneer, with $400 Billion to Hand Out." *New York Times*, May 11, 2023. https://www.nytimes.com/2023/05/11/climate/jigar-shah-climate-biden.html.

Prahalad, C. K., and Gary Hamel. "The Core Competence of the Corporation." *Harvard Business Review*, May–June 1990. https://hbr.org/1990/05/the-core-competence-of-the-corporation.

Priddle, Alisa. "GM Names Clegg New Production Chief for N. America." *Detroit Free Press*, June 24, 2014.

———. "VinFast Is Making Carmaking 101 Look Easy." *Motor Trend*, April 29, 2022. https://www.motortrend.com/news/vinfast-vietnam-factory-tour/.

Ramsey, Mike, and Gautham Nagesh. "LG Electronics to Be Key Supplier for GM Bolt." *Wall Street Journal*, October 20, 2015. http://www.wsj.com/articles/lg-electronics-to-be-key-supplier-for-gm-bolt-1445377888.

Reed, John. "GM Forms Partnership with Vietnam's VinFast." *Financial Times*, June 28, 2018. https://www.ft.com/content/a8da1eb4-7aa9-11e8-bc55-50daf11b720d.

———. "The Rise and Rise of a Vietnamese Corporate Empire." *Financial Times*, June 26, 2019. https://www.ft.com/content/84323c32-9799-11e9-9573-ee5cbb98ed36.

Rosevear, John. "Lucid Misses Revenue Expectations after EV Deliveries Disappoint." CNBC, August 7, 2023. https://www.cnbc.com/2023/08/07/lucid-lcid-q2-2023-earnings-guidance.html.

Scherr, Elana. "VinFast Is Going All-Electric, and Worldwide, at Full Speed." *Car and Driver*, April 27, 2022. https://www.caranddriver.com/features/a39816283/vinfast-update-factory-tour-vf8-prototype-drive/.

Schneider, Stefan, Siegfried Bajohr, Frank Graf, and Thomas Kolb. "State of the Art of Hydrogen Production Via Pyrolysis of Natural Gas." *Chemical & Biological Engineering Reviews* 7, no. 5 (July 13, 2020): 1–10.

Schneyder, Ernest. "GM to Help Lithium Americas Develop Nevada's Thacker Pass Mine." Reuters, January 31, 2023. https://www.reuters.com/markets/commodities/gm-lithium-americas-develop-thacker-pass-mine-nevada-2023-01-31/.

Selzer, Adam. "Burnham's 'Make No Little Plans' Quote: Apocryphal No More!" Mysterious Chicago, March 3, 2019. https://mysteriouschicago.com/finding-daniel-burnhams-no-little-plans-quote.

Smalley, Megan. "Liberty Tire Recycling, Bolder Industries Expand Partnership." *Recycling Today*, August 2, 2022. https://www.recyclingtoday.com/news/liberty-tire-recycling-bolder-formalize-partnership/.

Solis, Jeniffer. "Court Asked to Vacate NV Lithium Mine Approval After Precedent-Setting Case." *Nevada Current*, June 28, 2023. https://www.nevadacurrent.com/2023/06/28/court-asked-to-vacate-nv-lithium-mine-approval-after-precedent-setting-case/.

Sonner, Scott. "9th Circuit Denies Bid by Environmentalists and Tribes to Block Nevada Lithium Mine." Associated Press, July 17, 2023. https://apnews.com/article/nevada-thacker-pass-lithium-mine-4ad772a6940eb8edd507b50a179202f2.

"Spark EV" Wikipedia. Accessed August 7, 2023. https://en.wikipedia.org/w/index.php?title=Chevrolet_Spark&oldid=1185290384#Spark_EV.

St. Clair, Duane. "Honda Site Report Starts State Probe." *Columbus Dispatch*, September 28, 1977.

———. "State Wrapping Up Honda Negotiations." *Columbus Dispatch*, September 27, 1977.

Stanton, Shelby. *The Rise and Fall of an American Army: U.S. Ground Forces in Vietnam, 1963–1973.* New York: Random House, 2007.

"Statement by President Biden on Electric Vehicle and Battery Manufacturing Investments in North Carolina." The White House, March 29, 2022. https://www.whitehouse.gov/briefing-room/statements-releases/2022/03/29/statement-by-president-biden-on-electric-vehicle-and-battery-manufacturing-investments-in-north-carolina/.

Stevens, William. "Gas Stations: Tuning Up for the 1980s." *New York Times*, February 7, 1982, 4.

Stokes, Leah C. "What Joe Manchin Cost Us." *New York Times*, July 16, 2022. https://www.nytimes.com/2022/07/16/opinion/environment/joe-manchin-climate-change-clean-energy.html.

Stradling, Richard. "Photographer Who Took Iconic Vietnam War Photo Shoots Pics at VinFast Groundbreaking." *Raleigh News & Observer*, July 28, 2023. https://www.newsobserver.com/news/business/article277745313.html.

"Sun Group (Vietnam)." Wikipedia. Accessed July 8, 2023. https://en.wikipedia.org/wiki/Sun_Group_(Vietnam).

Talhelm, Matt. "Look Inside VinFast's Vietnam SUV Plant and How It Compares to the One Coming to NC." WRAL *TechWire*, July 21, 2022. https://wraltechwire.com/2022/07/21/look-inside-vinfasts-vietnam-suv-plant-and-how-it-compares-to-the-one-coming-to-nc/.

"Tesla Logo: Meaning, PNG, and Transparent Logo." CarLogos, May 19, 2023. https://www.carlogos.org/car-brands/tesla-logo.html.

Thomas, David. "Checking Out the Tesla Electric Roadster." Cars.com, December 1, 2006. https://www.cars.com/articles/checking-out-the-tesla-electric-roadster-1420663326614/.

Thompson, Clive. "When Coal First Arrived, Americans Said 'No Thanks.'" *Smithsonian Magazine*, July–August 2022. https://www.smithsonianmag.com/innovation/americans-hated-coal-180980342/.

Tobin, John. 1985. "Henry Ford and His Village Industries in Southeastern Michigan." Unpublished thesis, Eastern Michigan University, 1985.

Tracy, David. "Here's Why Tires Are Black." *Jalopnik*, November 15, 2018. https://jalopnik.com/heres-why-tires-are-black-1828925177.

"2024 Lucid Air Review, Pricing, and Specs." *Car and Driver*, August 8, 2023. https://www.caranddriver.com/lucid-motors/air.

"U.S. Automobile Production Figures." Wikipedia. Accessed July 6, 2023. https://en.wikipedia.org/wiki/U.S._Automobile_Production_Figures.

"US Electricity Profile 2022." U.S. Energy Information Administration. November 2, 2023. https://www.eia.gov/electricity/state/index.php.

Vance, Ashlee. *Elon Musk: Tesla, SpaceX, and the Quest for a Fantastic Future*. New York: HarperCollins, 2015.

Vanhulle, Lindsay. "GMC to Keep Hummer EV Customers Engaged as SUV Sales Begin." *Automotive News*, March 23, 2023. https://www.autonews.com/cars-concepts/gmc-hummer-ev-demand-growing-even-while-reservations-hold.

"Village industries." Wikipedia. Accessed August 13, 2023. https://en.wikipedia.org/wiki/Village_industries.

"VinFast Delays US Electric Vehicle Plant Operation to 2025." Reuters, March 10, 2023. https://www.reuters.com/business/autos-transportation/vinfast-delays-us-electric-vehicle-plant-operation-2025-2023-03-10/.

"VinFast Officially Delivers First VF 8 City Edition Vehicles to U.S. Customers | VinFast.," last modified March 1, https://vinfastauto.us/newsroom/press-release/vinfast-officially-delivers-first-vf-8-city-edition-vehicles-to-us-customers.

"Volkswagen Questioned by U.S. Senator Over Human Rights Issues Linked to China Suppliers." *Automotive News*, April 28, 2022. https://www.autonews.com/china/volkswagen-questioned-us-sen-marco-rubio-over-human-rights-issues-linked-china-suppliers.

Wayland, Michael. "Why GM Is Killing the Chevy Bolt—America's Cheapest EV—amid Record Sales." CNBC, April 29, 2023. https://www.cnbc.com/2023/04/29/why-gm-is-killing-the-chevy-bolt-ev-amid-record-sales.html.

———. "Why You Won't See Many Car Ads during Sunday's Super Bowl." CNBC, February 15, 2023. https://www.cnbc.com/2023/02/10/gm-jeep-kia-super-bowl-ads.html.

Wen Wei. "Cadillac's First LYRIQ EV Rolls Off Assembly Line in Shanghai." *Yicai Global*, May 9, 2022. https://www.yicaiglobal.com/news/cadillac-first-lyriq-ev-rolls-off-assembly-line-in-shanghai.

White, Joe. "VinFast: Party Like It's 2021." Reuters, December 7, 2022.

———. "GM Buys Cruise Automation to Speed Self-Driving Car Strategy." Reuters, March 11, 2016. https://www.reuters.com/article/us-gm-cruiseautomation-idUSKCN0WD1ND.

White, Joseph, and David Shepardson. "UAW Reaches Deal with GM, Ending Strike against Detroit Automakers." Reuters, October 30, 2023. https://www.reuters.com/business/autos-transportation/gm-reaches-tentative-deal-with-uaw-source-says-2023-10-30/.

Williams, Kevin. "'I'm Out': Ford F-150 Lightning Price Hikes Are Costing It Customers." *Verge*, June 27, 2023. https://www.theverge.com/2023/6/27/23771407/ford-f150-lightning-price-increase-cancel-order-dealer.

Williams, Mark. "Honda Battery Plant Helps Ohio Maintain Edge in Changing Auto Industry." *Columbus Dispatch*, October 16, 2022. https://www.dispatch.com/story/business/automotive/2022/10/16/honda-ohio-battery-plant-electric-vehicles-automotive-industry-economy/69558828007/.

———. "Meeting the Challenge: Ohio Report Outlines the Challenge Ahead in Move to EVs." *Columbus Dispatch*, July 24, 2023.

"Wind Market Reports: 2022 Edition." U.S. Department of Energy. Accessed August 10, 2023. https://www.energy.gov/eere/wind/wind-market-reports-2022-edition#wind Energy.gov., https://www.energy.gov/eere/wind/wind-market-reports-2022-edition.

Wood, Brayden. "Why Hydrogen Electric Vehicles Will be the Future of Auto Industry." TopSpeed, June 26, https://www.topspeed.com/hydrogen-electric-vehicles-future-of-auto-industry/.

Yang, Zeyi. "How Did China Come to Dominate the World of Electric Cars?" *MIT Technology Review*, February 21, 2023, https://www.technologyreview.com/2023/02/21/1068880/how-did-china-dominate-electric-cars-policy/.

Index

About the Author

Kenneth K. Boyer is Fisher Designated Professor of Operations and Business Analytics at Fisher College of Business, Ohio State University. He was the coeditor in chief of the *Journal of Operations Management* from 2006 to 2011. Previously Ken was a full professor at Michigan State University as well as on the faculty at DePaul University. His research interests focus on health care operations for patient safety, strategic management of operations, and the effective management of emerging technologies. He has memberships in the Academy of Management, Production and Operations Management Society, and the Decision Sciences Institute. Prof. Boyer has published over fifty academic articles, has been cited more than eleven thousand times and has written two books: *Operations and Supply Chain Management for the 21st Century* (2009) and *Extending the Supply Chain: How Cutting-Edge Companies Bridge the Critical Last Mile into Customers' Home*s (2004).